Demystifying Ansible Automation Platform

A definitive way to manage Ansible Automation Platform and Ansible Tower

Sean Sullivan

BIRMINGHAM—MUMBAI

Demystifying Ansible Automation Platform

Group Product Manager: Rahul Nair

Publishing Product Manager: Preet Ahuja

Senior Editor: Shazeen Iqbal

Content Development Editor: Romy Dias

Technical Editor: Nithik Cheruvakodan

Copy Editor: Safis Editing

Project Coordinator: Ashwin Dinesh Kharwa

Proofreader: Safis Editing

Indexer: Hemangini Bari

Production Designer: Joshua Misquitta

Marketing Coordinator: Nimisha Dua

First published: September 2022

Production reference: 1090922

Published by Packt Publishing Ltd.
Livery Place
35 Livery Street
Birmingham
B3 2PB, UK.

ISBN 978-1-80324-488-4

www.packt.com

This book is dedicated to my family, friends, and coworkers. I could not have done this without your help.

Contributors

About the author

Sean Sullivan works as a consultant for Red Hat and attended Purdue University, where he focused on networking. He has helped both small and large companies manage their Ansible deployments for the past five years, specializing in configuration as code and Ansible Automation Platform. He is a keen contributor to Red Hat's GitHub repository and an avid board gamer whose favorite game is Brass: Birmingham.

I want to thank those who have helped me along the way. Thank you to my parents, grandparents, and the rest of my family, who have always encouraged me to follow my path in life. Thank you to my friends, who are always there for me. Thank you to Tom Page, who has helped as a technical reviewer in this book and helps to maintain the redhat_cop collections with me. Thank you to Bianca Henderson and John Westcott, who put up with me and were instrumental in my understanding of the awx.awx collection modules. Thank you to David Danielsson, for all your help maintaining and contributing to the redhat_cop collections. Thank you to Kedar Kulkarni, for your help in starting the controller configuration.

About the reviewer

Tom Page has worked for Red Hat as a consultant for the past five years, with a key focus on working with customers to implement Ansible automation solutions. He is a co-maintainer of several AAP collections produced by Red Hat's Community of Practice, with the aim of delivering configuration as code as an end-to-end approach.

Table of Contents

3

Installing Ansible Automation Platform on OpenShift   39

4

Configuring Settings and Authentication   55

Part 2: Configuring AAP

5

Configuring the Basics after Installation   83

6

Configuring Role-Based Access Control   103

9

Automation Hub Management   167

10

Creating Job Templates and Workflows   185

Part 3: Extending Ansible Tower

11

Creating Advanced Workflows and Jobs   215

12

Using CI/CD to Interact with Automation Controller   229

13

Integration with Other Services   239

14

Automating at Scale with Automation Mesh   251

15

Preface

A few years ago during college, a friend introduced me to Ansible. I was taking networking classes at the time, and it opened my eyes to the future of automation. I was able to take a six-week lab that was coordinated between teams and simplify it as an Ansible deployment. Since then, I've spent years learning the ins and outs of using Ansible and **Ansible Automation Platform (AAP)**.

This book focuses on AAP and its many parts. The most popular part is the Automation controller, previously known as Tower. This book covers the different parts of AAP and how to utilize them. We'll focus on the Automation controller, which centralizes everything needed to run an Ansible playbook, using role-based access control, logging, and other features that are useful when doing automation in an enterprise.

In addition to the Automation controller, this book covers the **Automation hub**, the **Automation services catalog**, and **Red Hat Single Sign-On** and how they integrate and interact with the other services.

Also in this book is a collection of best practices around AAP. We will cover a variety of ways to interact with AAP. If you've ever had any interest in going beyond the command line of Ansible and scaling your automation, this book is a great first step in understanding AAP on that journey.

> Disclaimer
> The information and opinions provided in this work are solely those of the author and do not reflect the views of his employer.

Who this book is for

Many people I've encountered are familiar with Ansible or are able to get up to speed with Ansible quickly. However, few have taken to the idea of automating their automation. Most have written some playbooks and dabbled with Ansible Tower. Even if you have been using Ansible and parts of AAP for years, you will likely find new knowledge in this book.

If you are interested in diving into the nuances and stepping up to an intermediate-to-expert level in your understanding of Ansible Tower and AAP to integrate them into your infrastructure, then this book is for you.

In addition, there are multiple ways of interacting with the platform, including manually interacting with the web interface and using Ansible modules and roles to configure different parts of the platform. These roles can be used to implement **Configuration as Code (CaC)**.

What this book covers

Chapter 1, Introduction to Ansible Automation Platform, is an introduction to the AAP. It goes over the details of the different services that make up the platform and their upstream counterparts. In addition, this chapter goes into detail about the various methods used in this book and how to use execution environments.

Chapter 2, Installing Ansible Automation Platform, covers the installation of AAP on servers. This covers planning and scoping an installation, how to install Galaxy NG on a machine, how to backup and restore an installation, and a brief overview of troubleshooting installations.

Chapter 3, Installing Ansible Automation Platform on OpenShift, explores the installation of AAP with operators on an OpenShift cluster. This includes installing AWX on minikube in a development setup, installing the Red Hat OpenShift Local (formerly, CodeReady Containers) service for development testing, and how to backup and restore installations on OpenShift.

Chapter 4, Configuring Settings and Authentication, goes into detail about configuring automation controller settings, Red Hat Single Sign-On server configuration, integrating **Lightweight Directory Access Protocol** (**LDAP**) from Microsoft AD, how to add users and teams to the Automation controller without LDAP, and how to add users and groups to the Automation Hub without an identity provider.

Chapter 5, Configuring the Basics after Installation, deals with the basics of creating organizations, creating and configuring credential types and credentials, and exporting configurations from the automation controller.

Chapter 6, Configuring Role-Based Access Control, focuses on configuring role-based access control settings for the automation controller and Automation Hub. This involves setting various permissions for all objects and types in the AAP.

Chapter 7, Creating Inventory, and Other Inventory Pieces, covers creating and configuring inventories, and using and configuring inventory sources. It also includes an overview of popular inventory plugins, the use of inventory plugins from collections, and goes into detail about creating custom inventory plugins.

Chapter 8, Creating Execution Environments, looks at what **execution environments** (**EEs**) are, how they are used, how to create and modify your own EE, and how to use roles to create EEs using CaC.

Chapter 9, Automation Hub Management, delves into Automation Hub management. This includes going into detail about different content sources, how to set up and synchronize certified and community collections, publishing custom collections, managing execution environments and registries, and connecting the Automation Hub to the automation controller.

Chapter 10, Creating Job Templates and Workflows, goes into detail about creating projects, job templates, and workflows. It also goes into detail about using surveys with job templates and workflows, workflow creation using modules and roles, and how to utilize job slicing.

Chapter 11, *Creating Advanced Workflows and Jobs*, goes into detail about designing playbooks and jobs in a workflow to take advantage of workflows. This includes creating nodes that contain information so that users do not need to hunt in a playbook, and using approval nodes to gain user input to allow a workflow to continue. Additionally, this chapter will go into detail about creating and using notifications.

Chapter 12, *Using CI/CD to Interact with Automation Controller*, is all about using **continuous integration/continuous deployment** (**CI/CD**) and Webhooks. This goes into detail about maintaining AAP with CaC with CI/CD, how to launch jobs, monitoring and interacting with workflows, how to launch ad hoc commands, and the use of backup and restore for the AAP using CI/CD.

Chapter 13, *Integration with Other Services*, focuses on integration with other services, such as Red Hat Insights, Splunk, and Prometheus. These different services each focus on a different aspect of information generated by the platform. This includes job and event logs and metrics. Having a searchable log of this information can be invaluable for detecting problems and trends as your automation grows.

Chapter 14, *Automating at Scale with Automation Mesh*, teaches you how to automate at scale with automation mesh. This chapter goes into detail about using instance groups to set where automation runs, the use of different node types to scale AAP, and how to design an installation that works across disparate networks.

Chapter 15, *Using Automation Services Catalog*, explores what exactly the Automation services catalog is, how to configure the various parts of it, how to create orders and products for users to consume, and how to use approval workflows.

To get the most out of this book

This book assumes you know the basics of Ansible and are able to install and use it on your preferred operating system. In addition, as this book covers AAP, it assumes that you are able to set up one of the following virtual machines, OpenShift, or a minikube instance.

Software/hardware covered in the book	Operating system requirements
Ansible 2.13	Windows, macOS, or Linux
Python 3	Windows, macOS, or Linux
Docker/Podman	Windows, macOS, or Linux
Ansible Automation Platform 2.2	Windows, macOS, or Linux

The majority of the chapters assume you are using AAP. Instructions are provided on how to get a limited trial license for use in following along with the book, and there are also some instructions on how to install the developer versions of some software.

If you are using the digital version of this book, we advise you to type the code yourself or access the code from the book's GitHub repository (a link is available in the next section). Doing so will help you avoid any potential errors related to the copying and pasting of code.

If you have found any issues or corrections, or if you have any suggestions or other feedback, feel free to open an issue in this book's repository! I will strive to be present there and to maintain and update the code provided so that it continues to be useful, working, and accurate.

Download the example code files

You can download the example code files for this book from GitHub at `https://github.com/PacktPublishing/Demystifying-Ansible-Automation-Platform`. If there's an update to the code, it will be updated in the GitHub repository.

We also have other code bundles from our rich catalog of books and videos available at `https://github.com/PacktPublishing/`. Check them out!

Download the color images

We also provide a PDF file that has color images of the screenshots and diagrams used in this book. You can download it here: `https://packt.link/USfpC`.

Conventions used

There are a number of text conventions used throughout this book.

`Code in text`: Indicates code words in text, database table names, folder names, filenames, file extensions, pathnames, dummy URLs, user input, and Twitter handles. Here is an example: "Each list item can contain options, such as `extra_data` and `job_tags` listed in the previous section, but additional fields are also used."

A block of code is set as follows:

```
workflow_nodes:
  - identifier: Inventory Update
    related:
    unified_job_template:
    all_parents_must_converge: false
    extra_data: {}
```

When we wish to draw your attention to a particular part of a code block, the relevant lines or items are set in bold:

```
related:
  always_nodes: []
```

```
credentials: []
failure_nodes: []
success_nodes:
  - identifier: Template 1
```

Any command-line input or output is written as follows:

```
$ ansible-galaxy collection install awx.awx redhat_cop.
controller_configuration
$ ansible-navigator run demo.yml -m stdout
```

Bold: Indicates a new term, an important word, or words that you see onscreen. For instance, words in menus or dialog boxes appear in **bold**. Here is an example: "Select **System info** from the **Administration** panel."

Tips or Important Notes

Appear like this.

Current published book version information

This book references Ansible and several collections in Galaxy. As is bound to happen, updates and improvements will be made. For reference, these are the versions this book uses at the time of publication:

- Current Ansible Core version at the time of publication: 2.13

- Current AAP version at the time of publication: 2.2

- Current awx.awx collection version at the time of publication: 21.3.0

- Current redhat_cop.controller_configuration version at the time of publication: 2.1.6

- Current redhat_cop.ah_configuration version at the time of publication: 0.8.1

- Current redhat_cop.aap_utilities version at the time of publication: 2.1.0

- Current redhat_cop.ee_utilities version at the time of publication: 1.0.2

Get in touch

Feedback from our readers is always welcome.

General feedback: If you have questions about any aspect of this book, email us at `customercare@packtpub.com` and mention the book title in the subject of your message.

Errata: Although we have taken every care to ensure the accuracy of our content, mistakes do happen. If you have found a mistake in this book, we would be grateful if you would report this to us. Please visit `www.packtpub.com/support/errata` and fill in the form.

Piracy: If you come across any illegal copies of our works in any form on the internet, we would be grateful if you would provide us with the location address or website name. Please contact us at `copyright@packt.com` with a link to the material.

If you are interested in becoming an author: If there is a topic that you have expertise in and you are interested in either writing or contributing to a book, please visit `authors.packtpub.com`

Share your thoughts

Once you've read *Demystifying Ansible Automation Platform*, we'd love to hear your thoughts! Scan the QR code below to go straight to the Amazon review page for this book and share your feedback.

`https://packt.link/r/1803244887`

Your review is important to us and the tech community and will help us make sure we're delivering excellent quality content..

Part 1: Getting Ansible Automation Platform Up and Running

There are a few ways to get Ansible Automation Platform installed, and this section will go through the various installation options, their pros and cons, and some initial configuration.

The following chapters are included in this section:

- *Chapter 1, Introduction to Ansible Automation Platform*
- *Chapter 2, Installing Ansible Automation Platform*
- *Chapter 3, Installing Ansible Automation Platform on OpenShift*
- *Chapter 4, Configuring Settings and Authentication*

1

Introduction to Ansible Automation Platform

Ansible Automation Platform (AAP) has many parts, with its most popular part being the **Automation controller**, previously known as **Tower**. First, we will cover the different parts of AAP and how they interact with each other.

In addition, there are multiple ways of interacting with the platform, including manually interacting with the web interface, and using Ansible modules and roles to configure different parts of the platform. These roles can be used to implement **Configuration as Code (CaC)**.

In this chapter, we will cover the following topics:

- AAP overview
- Key differences between upstream and official Red Hat products
- Overview of the methods that will be used in this book
- Execution environments and Ansible Navigator

Technical requirements

In this chapter, we will cover the platform and methods that will be used in this book. In the *Overview of the methods that will be used in this book* section, the code that will be referenced can be found at `https://github.com/PacktPublishing/Demystifying-Ansible-Automation-Platform/tree/main/ch01`. It is assumed that you have Ansible installed to run the code provided. Additional Python packages will be referenced for installation.

AAP overview

This book will mainly focus on the Ansible Automation controller, which is the largest part of AAP. Most of these components have upstream projects that they relate to and allow you to report issues, look at the code, and even contribute.

The platform is made up of the following parts:

- Automation controller (formerly Red Hat Ansible Tower)
- Automation execution environments
- Automation hub
- Automation services catalog
- Red Hat Insights for Red Hat AAP
- Ansible content tools

This chapter will provide a brief description of each of these parts and how they relate to the Automation controller. A model of these relationships can be seen in the following diagram:

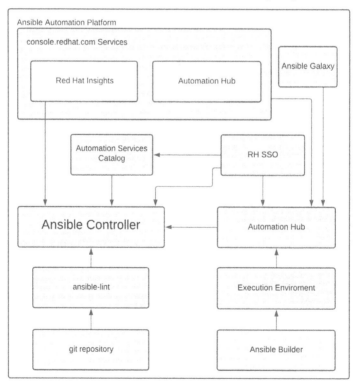

Figure 1.1 – AAP relationship model

In the next section, we will provide an overview of AAP, as shown in the preceding diagram.

Automation controller (Red Hat Ansible Tower)

The Automation controller is the workhorse of AAP. It is the central place for users to run their Ansible automation. It provides a GUI, **Role-Based Access Control** (**RBAC**), and an API. It allows you to scale and schedule your automation, log events, link together playbooks in workflows, and much more. Later in this book, we will look at the functions of the controller in more detail.

Up until recently, the Automation controller was referred to as Ansible Tower until it was split to be the Automation controller and various execution environments. This removed the reliance on Python virtual environments and allowed the platform and the command line to use execution containers.

> **Note**
>
> This book will refer to the Ansible Automation controller, though 90% of the time, it will still apply to the older versions of Tower.

But what does the Automation controller have to do with Ansible? The short answer is that the controller runs Ansible. It provides an environment that allows jobs to be repeated in an idempotent fashion – in other words, a repeatable way that does not change the final state. It does so by storing inventories, credentials, jobs, and other related objects in a centralized place. Add in RBAC, credential management, logging, schedules, and notifications and you can execute Ansible so that it's more like a service, rather than a script.

The upstream version of the Automation controller is AWX. It can be found at `https://galaxy.ansible.com/awx/awx`. AWX is an unsupported version of the controller and was built to be installed inside a Kubernetes cluster of some kind. It is frequently updated, and upgrades between versions are not tested. Fixes are not backported to previous AWX versions as they are with the Automation controller and Tower. For these reasons, it's recommended for enterprise users to use the Automation controller.

The Automation controller will pull collections and roles from Ansible Galaxy or a configured Automation hub. It will also pull container images from an image repository or Automation hub if none are specified.

Automation execution environments

The Automation execution environments are something new that was introduced with AAP 2.0. Previously, Ansible Tower relied on Python virtual environments to execute everything. This was not portable and sometimes made development difficult. With the introduction of execution environments, Ansible can now be run inside portable containers called execution environments. These containers are built from base images provided by Ansible that can be customized. Details of that customization will be covered in *Chapter 8, Creating Execution Environments*.

Execution environments can be built using the ansible-builder CLI tool, which takes definition files to describe what to add to a base image. They then can be executed using the ansible-navigator CLI/TUI tool to execute local playbooks. These same containers can be used inside the Ansible controller to execute playbooks as well, decreasing the difference between executing something locally and on the controller.

Automation hub

The Automation hub is an on-premises deployment of Ansible Galaxy (`https://galaxy.ansible.com/`). It can host collections and container images. It can be configured to get certified collections from Red Hat's Content Collections of certified content that reside on their Automation hub at `https://console.redhat.com/ansible/automation-hub`, from the public Ansible Galaxy, or any valid collection that's been uploaded by an administrator. It is also a place for users to go to discover collections that have been created by other groups inside an organization. The upstream repository of the Automation hub is `Galaxy_ng` (`https://github.com/ansible/galaxy_ng`). It is based on pulp and has much of the same features as the Automation hub.

Collections have replaced roles in terms of a form of distributing content. They contain playbooks, roles, modules, and plugins. They should be used so that code can be reused across playbooks. Automation hub was built so that users could store and curate collections on-premises without resorting to storing tarballs in a Git repository or a direct connection to Ansible Galaxy.

When you're installing Automation hub, it is possible to set up an image repository as well to host execution environment container images. This is not present when you're using the OpenShift operator as it is assumed that if you have an OpenShift cluster, you should already have an image repository.

Automation services catalog

The Automation services catalog provides a frontend for users to run automation using a simplified GUI that is separate from the Ansible Automation controller. It allows for multiple controllers to be linked to it, and for users to order *products* that will launch jobs. It also allows users to approve jobs through the services catalog. It is an extension of the Automation controller. A good example of this service can be found at `https://www.youtube.com/watch?v=Ry_ZW78XYc0`.

Red Hat Insights for Red Hat AAP

Red Hat Insights provides a dashboard that contains health information and statistics about jobs and hosts. It can also calculate savings from automation to create reports. Go to the following website to access the Insights dashboard: `https://console.redhat.com/ansible/ansible-dashboard`.

This includes the following:

- Monitoring the Automation controller's cluster health
- Historical job success/failure over time
- An automation calculator to approximate time and money that's been saved by using automation

Ansible content tools

While not directly related to the Ansible controller, Ansible has tools that assist in creating and developing playbooks and execution environments. These include the CLI/TUI tool known as ansible-navigator, which allows users to run playbooks in an execution environment, the ansible-builder CLI tool, which can be used to create execution environments, and `ansible-lint`, a linter that you can use to check your code to make sure it follows best practices and conventions, as well as identifying errors in code.

Another tool is the Ansible VS Code Extension for Visual Studio Code at `https://marketplace.visualstudio.com/items?itemName=redhat.ansible`. This is an IDE extension for syntax highlighting, keywords, module names, and module options. While there are several code editors out there, including Visual Studio Code, Atom, and PyCharm, to name a few, the Ansible Visual Studio Code Extension is a great way to double-check your Ansible work.

That concludes our overview of AAP. Now, let's address how this book goes about interacting with the different parts of the platform.

Key differences between upstream and official Red Hat products

Earlier, we briefly mentioned upstream projects. The key ones are AWX and `Galaxy_ng`. These projects are built to be bleeding edge regarding rapid changes as developers from both the public and Red Hat make changes and improvements. Things are expected to break, and the upgrade path from one version to another is not guaranteed or tested. Bug fixes are also *not* backported to previous versions. However, their downstream versions, such as the Automation controller and Automation hub, go through rigorous testing, including testing on upgrading from one version to another. Not all the changes that are made upstream make it to the next release of the downstream product. In addition, most bug fixes do get backported to previous versions.

For these reasons, it is not recommended to use upstream products in production. Because of these caveats, they are fine for a home lab, proof of concept, and development, but *not* production.

Overview of the methods that will be used in this book

This book is built around defining CaC. What this means is that along with showing you how to define something in the GUI, a more Ansible approach will be presented. Ansible is a method of automation, and time and time again I have seen folks manually creating and setting their automation through manual processes. In that vein, this will serve as a guide to automating your automation through code.

The benefits of CaC are as follows:

- Standardized settings across multiple instances, such as development and production.
- Version control is inherent to storing your configuration in a Git repository.
- Easy to track changes and troubleshoot problems caused by a change.
- Ability to use CI/CD processes to keep deployments up to date and prevent drift.

A best practice is that you create a development environment to make changes, initial deployments, and run tests. Nothing ever runs perfectly the first time – it is through iteration that things improve, including your automation. Through the methods described here, using this method helps prevent drift and allows you to keep multiple instances of an Automation controller configured and synced, such as development and production, which should be a simple process.

There are three approaches you can take to managing the services as part of AAP: the manual approach and managing the configuration through Ansible modules or roles.

Introduction to the roles and modules that will be used in this book

Throughout this book, various roles and modules will be referred to that belong to collections. Collections are a grouping of modules, roles, and plugins that can be used in Ansible.

`awx.awx`, `ansible.tower`, and `ansible.controller` are module collections that can be used interchangeably. These are all built off of the same code base. Each is built to be used with their respective product – that is, AWX, Tower, and the Automation controller.

`redhat_cop.controller_configuration` is a role-based collection. This means that it is a set of roles that's been built to use one of the three aforementioned collections to take definitions of objects and push their configurations to the Automation controller/Tower/AWX.

The `redhat_cop.ah_configuration` collection is built to manage the Automation hub. It contains a combination of modules and roles that are designed to manage and push configuration to the hub. It is built on the code from both of the previous collections, but specifically is tailored to the Automation hub.

`redhat_cop.ee_utilties` is built to help build execution environments. Its role is to help migrate from Tower to the Automation controller and build execution environments from definition variables.

The last one we will mention is `redhat_cop.aap_utilities`. This is a collection that was built to help with installing a backup and restore of the Automation controller and other useful tools that don't belong with the other controller collections.

The manual approach

Nearly everything after the installer can be set manually through the GUI. This involves navigating to the object page, making a change, and saving it, which works fine if you are only managing a handful of things, or making one small change. For example, to create an organization using the Ansible controller web interface, follow these steps:

1. Navigate to the Ansible controller web interface; for example, `https://10.0.0.1/`.

2. Log in using your username and password.

3. On the home page in the left-hand section, select **Organizations** | **Add**.

4. Fill in the name, description, and any other pertinent sections.

5. Click **Save**.

The following screenshot shows an example of the page where you can add an organization:

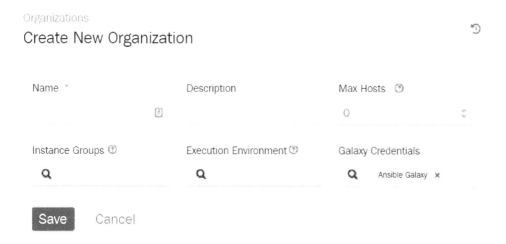

Figure 1.2 – Create New Organization

This method can be repeated for all the objects in the Automation controller. Although it is prone to mistakes, it can be useful for making quick changes when you're testing before committing the changes to code.

Using Ansible to manage the configuration

The best method is using Ansible to automate and define your deployment. The Ansible team has created a collection of modules that you can use to interact with the Automation controller. Upstream, this is known as `awx.awx`, and the official collection is named `ansible.controller`. The code is roughly the same between the two, but the latter has gone through additional testing and is supported by Red Hat. `ansible-galaxy` will need to be used to install the collections. You can use either of the two commands to do so:

```
ansible-galaxy collection install awx.awx redhat_cop.
controller_configuration
ansible-galaxy collection install -r requirements.yml
```

This file can be found in `/ch01/requirements.yml` in this book's GitHub repository: `https://github.com/PacktPublishing/Demystifying-Ansible-Automation-Platform`.

There is a module for each object in the Automation controller. For example, the following module is used to create an organization:

```
// create_organization_using_module.yml
---
- name: Create Organization
  hosts: localhost
  connection: local
  gather_facts: false
  collections:
    - ansible.controller
  tasks:
    - name: Create Organization
      ansible.controller.organization:
        name: Satellite
        controller_host: https://10.0.0.1
        controller_username: admin
        controller_password: password
        validate_certs: false
  ...
```

These modules do the heavy lifting of finding object IDs and creating the related links between various objects. This especially simplifies the creation of an object such as a job template.

Alongside this collection of modules, consultants at Red Hat, and a few other people working with the Ansible Automation controller, came up with the redhat_cop.controller_configuration collection. A series of roles was created to wrap around either the awx.awx or ansible.controller collection, to make it easier to invoke the modules and define a controller instance, as well as several other collections to help manage other parts of AAP. This book will assume you are using one of these two collections in conjunction with the redhat_cop collections.

The basic idea of the controller configuration collection is to have a designated top-level variable to loop over and create each object in the controller. The following is an example of using the controller configuration collection:

```
// create_organization_using_role.yml
---
- name: Playbook to push organizations to controller
  hosts: localhost
  connection: local
  vars:
    controller_host: 10.0.0.1
    controller_username: admin
    controller_password: password
    controller_validate_certs: false
    controller_organizations:
      - name: Satellite
      - name: Default
  collections:
    - awx.awx
    - redhat_cop.controller_configuration
  roles:
    - redhat_cop.controller_configuration.organizations
...
```

This allows you to define the objects as variables, invoke the roles, and have everything created in the controller. It is often easier to import a folder full of variable files, such as the following task:

```
// create_objects_using_role_include_files.yaml
---
- name: Playbook to push objects to controller
  hosts: localhost
  connection: local
  collections:
    - awx.awx
```

```
         - redhat_cop.controller_configuration
     pre_tasks:
       - name: Include vars from configs directory
         include_vars:
           dir: ./configs
           extensions: ["yml"]
     roles:
         - redhat_cop.controller_configuration.organizations
         - redhat_cop.controller_configuration.projects
  . . .
```

The included files define the organizations exactly like the previous task, but the projects are defined as follows:

```
configs/projects.yaml
---
controller_projects:
  - name: Test Project
    scm_type: git
    scm_url: https://github.com/ansible/tower-example.git
    scm_branch: master
    scm_clean: true
    description: Test Project 1
    organization: Default
    wait: true
    update: true
  - name: Test Project 2
    scm_type: git
    scm_url: https://github.com/ansible/ansible-examples.git
    description: Test Project 2
    organization: Default
  . . .
```

Because the GUI, modules, and roles are all the same information in different forms, each section will contain details about creating the YAML definitions and how to use them.

Using these methods is the primary way to interact with AAP as a whole. The focus will be on CaC since it is the recommended way of interacting with the Automation services. Like most tasks in Ansible, it is one of many ways to do the same thing.

Execution environments and Ansible Navigator

A newer feature of Ansible is the addition of execution environments. These are prebuilt containers that are made to run Ansible playbooks. These replace using Python virtual environments as the standard way of using different versions of Python and Python packages. The Ansible controller takes advantage of these environments to scale and run job templates as well. They solve the issue of *it works for me*, maintaining different environments across all nodes, and other problems that arose from the previous solution. They also double as a simplified developmental version of the Automation controller for testing a job template when you're using the same container as the controller:

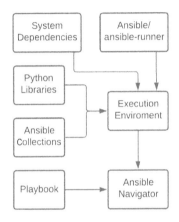

Figure 1.3 – Ansible Navigator inputs

`ansible-navigator` was built to replace `ansible-playbook`; it allows you to run playbooks in a container in the command line, similar to how jobs are run inside the controller. To install `ansible-navigator`, use the `pip3 install 'ansible-navigator[ansible-core]'` command on your desired machine. Afterward, you can run the `demo.yml` playbook in the `ch01` folder:

```
//demo.yml
---
- name: Demo Playbook
  hosts: localhost
  gather_facts: false
  tasks:
    - debug:
        msg: Hello world
...
```

To run this playbook in a container, use the `ansible-navigator run demo.yml -m stdout` command. It should output a `Hello world` message. Using the `-ee` or `-eei` option, the execution environment can be specified. This allows the user to use the same execution environment that was used in the controller for testing and development.

Additional Python libraries and collections can be added to an execution environment, which will be covered in *Chapter 8*, *Creating Execution Environments*. Additional information can also be found at `https://ansible-builder.readthedocs.io/en/stable/`.

Summary

Now that you know how AAP interacts with services and have been introduced to the methods that will be used in this book, you are armed with the information you need to go further.

In the next chapter, we will cover how to install the controller and AAP on physical machines.

Installing Ansible Automation Platform

There are a few different ways to install **Ansible Automation Platform** (**AAP**). The two most popular are installing them on machines (either physical or virtual) or an OpenShift cluster. In addition, there are upstream options for installing the latest bleeding-edge versions of the Automation controller and Automation hub.

In this chapter, we will cover the following topics:

- Planning an installation on a machine
- Installing Galaxy NG on a machine
- Backing up and restoring the installation
- Issues that can arise when using the installer

Technical requirements

This chapter will go over the platform and methods used in this book. The code referenced in this chapter is available at `https://github.com/PacktPublishing/Demystifying-Ansible-Automation-Platform/tree/main/ch02`. It is assumed that you have Ansible installed to run the code provided.

In addition, it is assumed that you have either one to five virtual machines or physical machines that have **Red Hat Enterprise Linux** (**RHEL**) installed to install the different parts of AAP. These servers are required to follow the instructions provided, and it is highly recommended to at least follow *one* of the methods from either this chapter or *Chapter 3, Installing Ansible Automation Platform on OpenShift*, to complete the rest of the chapters in this book.

Getting a trial version of Ansible Automation Platform

A trial version of AAP can be obtained with a Red Hat Developer account. To get a Developer account, register at `developers.redhat.com`. After registering for a Developer account, navigate to `https://developers.redhat.com/products/ansible/getting-started` and download the installer to start your trial subscription. Once the subscription is activated, it will populate a subscription at `access.redhat.com/management/subscriptions`. This subscription license will be needed later in this chapter, and in *Chapter 4, Configuring Settings and Authentication*, when we set up an Automation controller. This is not needed if you choose to use AWX and Galaxy NG.

Planning an installation on a machine

There are two ways to install AAP. The first involves using servers, with the latter taking advantage of OpenShift clusters. The server can be installed using physical machines or any form of **virtual machine** (**VM**). Generally, it is recommended to use a virtual machine of some kind, be it a cloud instance, something on OpenStack, or even a VM that is running on your laptop. It is assumed you know how to create these instances, as their creation and maintenance are well beyond the scope of this book. For simplicity, the term *node* will be used to describe a virtual machine instance.

At a minimum, the Ansible Automation controller requires a single node. That node will host the database, the control plane, and the execution portion. The same thing can be said for Automation hub. It can be installed separately on a single node, hosting both the hub and the database it uses. Both the Automation controller and Automation hub must be installed on separate nodes as they both have a web interface. Single-node deployments of each are not recommended, but when creating a development environment, something that isn't used in production, or something as a proof of concept, they are still valid deployments. An optimum minimal deployment for an enterprise would look like this:

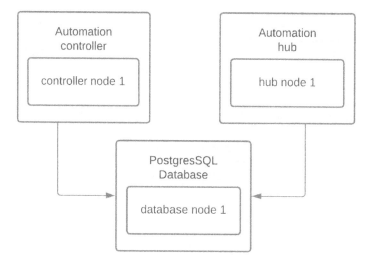

Figure 2.1 – Ansible Automation Platform minimal deployment

It is recommended to not use the option where the database and the service are on the same node. This leads to problems if you want to grow the deployment by adding more Automation controllers or Automation hub nodes. This minimal design allows for expansion and growth as needed.

The distinct types of nodes

In the past, Ansible Tower required a minimum of one or three nodes so that it would have a quorum when using RabbitMQ. However, newer versions of the controller do not have this requirement. A new feature called the automation mesh was introduced with the introduction of AAP 2.1. This allows for the separation of the control plane and the execution parts of the controller. It also allows for execution nodes to be on-premises, in the cloud, or even on the edge.

The separation of the control and execution planes adds options for deployments. It is recommended to have at least two to three control nodes, and as many execution nodes as needed to accommodate your peak number of jobs. Another option is to use instance groups to allow execution on an OpenShift deployment, an option that will be discussed in Chapter 3, *Installing Ansible Automation Platform on OpenShift*, which allows users and organizations to specify where they want to run their jobs and allows for hybrid server/OpenShift capacity to be used.

There are *eight* types of nodes in AAP deployment:

- **control node**: This only provides the web GUI and API
- **execution node**: This only executes jobs
- **hybrid node**: This combines a control and execution node
- **hop node**: Also known as a jump host, this node does not execute jobs but routes traffic to other nodes
- **database node**: This hosts the PostgreSQL database
- **automation hub node**: This hosts the pulp files and the GUI and API
- **sso_node**: This hosts the Red Hat Single Sign-On (SSO) server
- **catalog_node**: This hosts the Automation services catalog server

The minimum requirements are as follows. A service node is any type of control, database, execution, or hybrid node.

Node type	CPU core	RAM (GB)	Disk (GB)	IOPS
service	4	16	40	1500
hub	2	8	60	1500
hop / catalog	4	16	20	1500
sso	2	8	6	1500

Table 2.1 – Ansible Automation Platform node requirements

These are the minimum requirements, though more CPU cores and RAM can always be added. In practice, production deployments run three nodes with 16, 32, or more CPUs and memory ranging from 128 GB to 300 GB per node. For databases, 150 GB of disk space is recommended, but it can vary in size, depending on how many jobs and events are used each month. Every deployment will be different, so it is best to monitor what your cluster is using and adjust appropriately.

Sizing requirements depend on the number of hosts and the number of jobs run each day per host. There are a few factors that play into this. Any given instance can be restrained by memory or CPU-bound capacity. This capacity determines how many forks can be run concurrently.

Capacity is determined by the following factors:

- Memory-bound capacity:

  - (instance memory – 2,048)/memory per fork (set to 100)

  - Example: *(16,384-2,048)/100 = 143.36*

- CPU-bound capacity:

  - CPU cores * 4 = CPU capacity

  - Example: *4*4 =16*

This determines how many jobs can run at a time concurrently. By default, the instances are bound by memory. If you have automation running where there are a lot of jobs running, more memory or CPUs might be needed.

As a general rule of thumb, the following equations can be used to determine rough sizing recommendations:

- *Forks needed = (hosts * jobs per day per host * jobs duration for one host) / work hours*

  Example: *(2,000 hosts * 2 jobs * .15 duration) / 10 work hours = 60 forks*

- *Memory needed = ((Forks * memory per fork) +2,048 * nodes in an instance group) / 1,024*

  Example: *((60 forks * 100 memory per fork) + 2,048 * 4 instances)/1,024 = 14 GB*

- *CPUs needed = Forks needed by forks per CPU core*

  Example: *60/4 = 15*

If automation is spread to use more hours in a day and is scheduled, this can reduce the number of resources needed. The Automation controller will also monitor and queue jobs if it hits a capacity limit. It is best to monitor usage and adjust resources according to the needs of the business.

Another thing to take into account is the amount of disk space needed in the database. The following should be taken into account:

- Inventory storage is roughly 200 MB for 1,000 hosts. This accounts for the inventory and facts stored about the hosts.

- *Database size for jobs = hosts * jobs per day per host * tasks per playbook (events) * days to store (120) day default setting) * event size / 1,024.*

  Example: *(2,000 * 2 jobs per day * 30 tasks *120 day retention * 2 Kb) / 1,024 = 28,125 MB.*

  The days to store job events can be changed in the **Cleanup Job Schedule** management job on the Automation controller.

  The Red Hat certified collections, community collections, and published collections and execution environment images should be well under 1 GB of the database's size.

  Adding all of these together yields *(28,125 Job data +400 facts+1,024 hub)/1,024 = 28.85 GB* for the database.

The database should have a minimum of 1,500 **input/output operations per second** (**IOPS**) as well. The recommended size for the database machine is also 4 CPUs and 16 GB of RAM, but this should be monitored for bottlenecks under peak performance.

These are all theoretical and recommended resources for determining the initial size of a deployment. It is always recommended to use logging software to keep track of CPU, memory, and disk usage to determine whether you've over or under-estimated the actual usage in your deployment. The next section will go into detail about the automation mesh, which links instances together.

Automation mesh

With a combination of control, hop, and execution nodes, it is possible to set up what is referred to as a mesh of controller nodes. This allows for connections in disparate networks to use a combination of hop nodes and execution nodes.

The following diagram shows how to use a hop node with execution nodes in disparate networks:

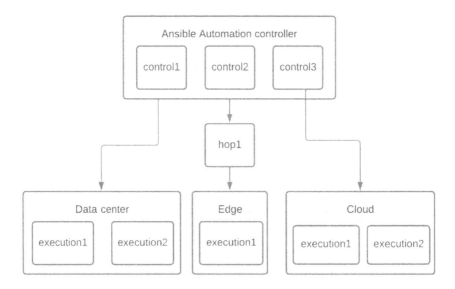

Figure 2.2 – Ansible Automation controller using hop and execution nodes

For firewall purposes, the following ports need to be opened for services:

Protocol	Port	Purpose	Nodes
SSH	22/TCP	AAP installation	All
HTTP/HTTPS	80/443/TCP	Web UI, API	Control, hybrid, hub, service catalog
HTTPS	443/TCP	Execution environment (EE) pulls	Execution, hybrid
Postgres	5432/TCP	Database connection	Control, non peered execution*
Receptor	27199/TCP*	Automation mesh	All
HTTP/HTTPS	8080/8443	SSO	SSO

Table 2.2 – Ansible Automation controller ports required

Non-peered execution nodes are those that are not peered back to the main cluster. Execution nodes in the main cluster will still need access to the PostgreSQL port. For reference, this means those in the Ansible Automation controller group in *Figure 2.2* need access, while the execution nodes in the data center, edge, and the cloud do not need a connection to the PostgreSQL server.

Hop nodes are specifically designed to require minimal ports open for use. They are used to manage or access devices in another security zone and require only a single port for the receptor to be open for communication.

In the *Installation methods* section, you will learn how to designate peers and hop nodes to implement the automation mesh.

High availability

Many users want high availability and load balancing for the services. A consideration for the Automation controller is that it is much more resource-dependent than its counterparts, which are Automation hub and the database. Databases can also be set up to be highly available to the services. However, PostgreSQL high availability can take on many forms and configurations. Each form depends on what priority you want to take in terms of availability. For this reason, we will discuss the node configurations for AAP to use an external database, though the discussion will not cover how to set one up outside the installer or how to configure high availability.

For the Automation controller, two to three hybrid nodes are recommended with a load balancer for a single point of access to the web GUI. For Automation hub, one to two nodes are recommended, with another load balancer to access it. As discussed previously, more nodes may need to be added, depending on the amount of capacity needed. Between the Automation controller and Automation hub, a single database is required, but this can also be split into separate database nodes. The following diagram illustrates a deployment with separate database nodes:

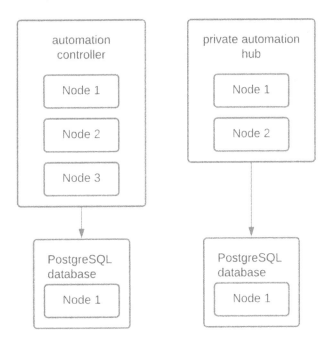

Figure 2.3 – Ansible Automation Platform recommended deployment

A high availability Automation hub requires a shared filesystem between the nodes. This can be achieved by using the following steps on all Automation hub nodes.

This can be done in many ways, though using NFS is the most common. To set up the shared filesystem, follow these steps:

1. When using a firewall, add a zone for the firewall using the following commands:

    ```
    $ sudo firewall-cmd --zone=public --add-service=nfs
    $ sudo firewall-cmd --zone=public --add-service=mountd
    $ sudo firewall-cmd --zone=public --add-service=rpc-bind
    ```

2. Reload and check that the firewall is enabled:

    ```
    $ sudo firewall-cmd –reload
    $ sudo firewall-cmd --get-services
    ```

    Once the firewall has been taken care of, make sure that the var/lib/pulp directory exists across all Automation hub nodes and enable the NFS share.

3. Make sure that the /var/lib/pulp directory exists by running the following command:

    ```
    $ sudo mkdir /var/lib/pulp
    ```

4. Enable the NFS share:

    ```
    $ sudo mount -t nfs4 <nfs_share_ip_address>:/ /var/lib/
    pulp
    ```

5. Ensure that the filesystem was mounted:

    ```
    $ df -h
    ```

Without this preparation step, the installer will fail when doing a multinode Automation hub installation.

Installation methods

This book will go over several methods of installing and maintaining pieces of AAP. The installers and applications themselves are built so that they can be configured and installed using just the tools Red Hat gives you. However, this book is guided toward defining all deployments with **Configuration as Code** (**CAC**) using machine-readable YAML definition files, and taking advantage of the provided roles and modules. The infrastructure can be tweaked, modified, maintained, and recreated by updating or using these files. If you are installing the upstream AWX component of AAP, please refer to *Chapter 3*, *Installing Ansible Automation Platform on OpenShift*.

> **Important Note**
>
> If you are using a VM, then make a snapshot of the VM before proceeding. The installer is very resilient, but it never hurts to have a backup to revert to.
>
> In addition, if you're upgrading an existing version of AAP, run a backup before running the installer. Backing up the installation will be covered later in this chapter.

Before we start with the installation, we need to download the installer ZIP file. This is available for AAP 2 and above at `https://access.redhat.com/downloads/content/480` and for 3.8.x at `https://releases.ansible.com/ansible-tower/`. Use the `tar -xvf` command on the downloaded file.

A base installer provides a blank inventory file. At the time of writing, the following is populated. The first section is the node groups and nodes to install the Automation controller on. The base install file assumes only the Automation controller has been installed on a single node, which also contains both the controller and the database:

default_inventory.ini

```
[automationcontroller]
localhostansible_connection=local

[automationcontroller:vars]
peers=execution_nodes

[execution_nodes]

[automationhub]
```

Note that the database node is blank. If this is left blank, the database will be installed on the `automationcontroller` node:

```
[database]

[servicescatalog_workers]

[sso]

[all:vars]
```

An admin password must be set; however, a default password is not supplied:

```
admin_password=''

pg_host=''
pg_port=''

pg_database='awx'
pg_username='awx'
pg_password=''
```

More variables are included in the blank inventory of the installer if you are only using a single-node Automation controller installation. These variables will be referenced in the next section.

The Ansible groups in the preceding INI inventory file are required when installing those specific services. For example, if the controller hosts group contains no nodes, then the installer will not install any controller nodes. It also contains the required variables for the installer. A breakdown of these variables is shown in the following tables.

Controller values

These are the base variables for the controller. Database variables must also be set:

Variables	Default Value	Possible Values	Notes
admin_password		password	Admin password for controller
receptor_listener_port	27199	27199	Port to use for receptor
generate_dot_file		File location	Where to put the graph dot file representation of mesh nodes

Table 2.3 – Ansible Automation controller base variables

PostgreSQL database values

These are the variables for the controller's PostgreSQL database. These are required for both Automation hub and the Automation controller:

Variables	Default Value	Possible Values	Notes
pg_host		Database FQDN	pg host to use
pg_port		5432	pg port to use
pg_database	awx	awx	pg database name
pg_username	awx	awx	pg database username
pg_password		password	pg database password
pg_sslmode	prefer	prefer | verify-full	Set to 'verify-full' for client-side enforced SSL

Table 2.4 – Ansible Automation controller PostgreSQL variables

Container registry values

The following variables are used to point to an external container registry. The default option is to authenticate and pull from the official Red Hat registry, but the default values can be changed to point to another registry if needed. This can be useful if the AAP is installed behind a firewall:

Variables	Default Value	Possible Values	Notes
registry_url	https://registry.redhat.io		URL for the container registry
registry_username		username	Redhat.com username or registery username
registry_password		password	Redhat.com password or registery password

Table 2.5 – Ansible Automation container registry variables

Automation hub values

The following values are set for Automation hub, including the attached database. If the database host is the same as the controller database, set the same host value:

Variables	Default Value	Possible Values	Notes
automationhub_admin_password		password	Automation hub password
automationhub_pg_host		database FQDN	pg host to use
automationhub_pg_port		5432	pg port to use
automationhub_pg_database	automationhub	automationhub	pg database name
automationhub_pg_username	automationhub	automationhub	pg database username
automationhub_pg_password		password	pg database password
automationhub_pg_sslmode	prefer	prefer | verify-full	Set to 'verify-full' for client-side enforced SSL
automationhub_require_content_approval	TRUE	True | False	Optional: All uploaded collections require approval
automationhub_ssl_validate_certs	TRUE	True | False	Optional: Validate certs on Automation Hub
automationhub_collection_signing_service_key			Absolute path to signing service key
automationhub_collection_signing_service_script			Absolute path to signing service script
automationhub_auto_sign_collections	FALSE	True | False	Sign collections by default

Table 2.6 – Ansible Automation hub variables

SSO values

The following are passwords used for the **Single Sign-On (SSO)** server node:

Variables	Default Value	Possible Values	Notes
sso_console_admin_password		secret123	Password' is too simple and will cause issues
sso_keystore_password		N/A	The password for the keystore to use for encryption

Table 2.7 – Ansible Automation single sign-on variables

Service Catalog values

The following table shows the values set for the Service Catalog, including the attached database. If the database host is the same as the controller database, set the same host value:

Variables	Default Value	Possible Values	Notes
automationcatalog_pg_host		database FQDN	pg host to use
automationcatalog_pg_port	5432	5432	pg port to use
automationcatalog_pg_database	automationservicescatalog	automationservicescatalog	pg database name
automationcatalog_pg_username	automationservicescatalog	automationservicescatalog	pg database username
automationcatalog_pg_password		secret123	pg database password
automationcatalog_controller_verify_ssl	FALSE	True | False	Whether or not to verify the SSL of the controller connection
sso_automationcatalog_create_user_group	TRUE	True | False	Whether to create default users and groups in the SSO

Table 2.8 – Ansible Automation Service Catalog variables

There are around 100 variables that can be overwritten when installing AAP; the ones shown in the preceding tables are the most common. To discover more, look in the installer and peruse the roles for the default variables that are contained inside `collections/ansible_collections/ ansible/automation_platform_installer/roles`. Specifically, look at the `<role>/ defaults/main.yml` file for each role's variables that can be overwritten.

Each of these variables is set in the inventory file of the installer. Once set, the installer can be invoked with the `./setup.sh` command. The output of the installer will be set to a log file in the `ansible- automation-platform-setup` folder. If any problems arise, investigate the log. Often, a specific variable will be missing, or an option in a specific task or module. Use the task name and `grep -rnw 'task name'` to find where in the roles the error is occurring to try and troubleshoot.

These are the basics for installing AAP. However, some tools have been written to ease this process – specifically, the `redhat_cop.aap_utilities` collection. This was built to download the installer file, extract it, create the inventory file from variables, and then run the setup. This requires an offline token to be generated, which you can do at `https://access.redhat.com/management/ api/`. Here is a sample playbook for invoking the roles:

aap_installation_basic.yml

```
---
- name: Playbook to install the AAP Platform
  hosts: localhost
  connection: local
  vars:
    aap_setup_down_offline_token: password123
    aap_setup_working_dir: /home/username/Documents/aap_lab_
prep
```

```
    aap_setup_down_type: setup
  vars_files:
    - inventory_vars/variables.yml
  roles:
    - redhat_cop.aap_utilities.aap_setup_download
    - redhat_cop.aap_utilities.aap_setup_prepare
    - redhat_cop.aap_utilities.aap_setup_install
...
```

> **Important Note**
> Back up your inventory file or variable reference! If you need to upgrade, back up, or restore
> the installation, the inventory file is key in that process.

One thing to notice is that a variable file was included in this playbook. A lot of variables are used in the installation, so to keep things clear, a separate variable file was used in the `inventory_vars` directory. The full file can be found in this book's GitHub repository (`inventory_vars/variables.yml`). Excerpts from the variable file are referenced as follows. The first top-level variable, `aap_setup_prep_inv_nodes`, covers each inventory group and a node in that group.

After each host, additional variables can be included for that particular host.

In this example, the variable is `servicecatalog_controller_name`:

```
  aap_setup_prep_inv_nodes:
    automationcontroller:
      controller.node:
    database:
      database
```

The following top-level variable, `aap_setup_prep_inv_vars`, covers all the other variables for the installer:

```
  aap_setup_prep_inv_vars:
```

The following top-level variable is a dictionary of dictionaries. The first level key (`automationcontroller`) is the inventory group name, while the second level key (`peers`) is the variable name with the value (`execution_nodes`):

```
    automationcontroller:
      peers: execution_nodes
      node_type: hybrid
```

```
execution_nodes:
  node_type: execution
```

All the variables after this are global:

```
all:
```

The following variables are the base variables needed for installation, which are set as dictionary-value pairs. The installation variables from *Tables 2.3-2.8* are used here:

```
admin_password: 'secret123'
pg_host: 'database'
pg_port: '5432'
pg_database: 'awx'
pg_username: 'awx'
pg_password: 'secret123'
pg_sslmode: 'prefer'  # set to 'verify-full' for client-
side enforced SSL
registry_url: 'registry.redhat.io'
receptor_listener_port: 27199
...
```

Note that the secrets are separate as they are built to be loaded in from a vaulted variable file or could be included inline with a vaulted variable. There are also host `vars` listed on the second line that can be added, such as `servicescatalog_controller_name`.

This configuration will install services on five nodes, as shown in the following diagram, with the full suite of automation services. The following diagram shows how these nodes interact. The key thing is that the SSO services integrate with Automation hub and the Automation controller's LDAP, and the Automation controller and Automation hub use the database. Removing the Ansible group from the installer will skip setting up that particular node:

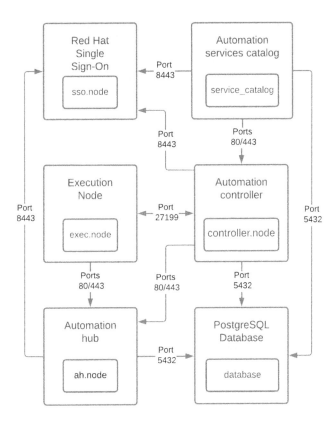

Figure 2.4 – Ansible Automation Platform Enterprise with all six services and their interactions

In addition, there is a variable called `generate_dot_file` that will generate a Graphviz notation file of the nodes in your inventory. For example, with a controller node and an execution node, it will generate the following file:

ch02/graphdot

```
strict digraph "" {
    rankdir = TB
    node [shape=box];
    subgraph cluster_0 {
        graph [label="Control Nodes", type=solid];
        {
            rank = same;
            "controller.node";
```

```
      }
    }
    "exec.node";
    "controller.node" -> "exec.node";
}
```

This file, when rendered on a GraphDot editor, such as `http://magjac.com/graphviz-visual-editor/`, will generate a graphical representation of the nodes:

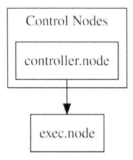

Figure 2.5 – GraphDot representation of nodes, as generated by the AAP installer

Setting variables for the automation mesh and node types

Automation mesh and node types are set with variables in the inventory file. The mesh is a replacement for isolated nodes from the previous 3.x version of Ansible Tower. Communication is determined by what nodes peer to each other. This allows for hop nodes and nodes in disparate networks. An example of remote nodes with and without a hop node is shown in the following diagram:

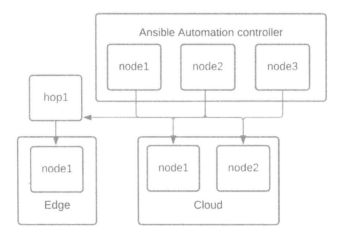

Figure 2.6 – Node representation for the Automation controller mesh

This installer takes the peers and node_types variables from the inventory and uses them to create a mesh. The node type determines what is installed on each node. By default, nodes in the execution_nodes group are execution nodes. The following snippet from the inventory file can be used to create the mesh shown in the preceding diagram:

mesh_snippet.ini

```
[automationcontroller]
node1
node2
node3

[automationcontroller:vars]
peers=hop1
node_type=control

[execution_nodes]
cloud1
cloud2
edge1

[hop_nodes]
hop1 node_type=hop peers=automationcontroller
[cloud_nodes]
cloud1
cloud2
[edge_nodes]
edge1 peers=hop_nodes
```

The automation mesh and node types can be customized to be as complicated or as simple as defined. Network requirements may require peers to be defined. It is up to the administrator to determine what is needed. *Table 2.2* will help you determine what is needed.

> **Important Note**
> As discussed previously, non-peered execution nodes require access to the PostgreSQL database. In this example, those are the cloud and the Ansible Automation controller group nodes. The edge node does not need to have a route to the PostgreSQL database.

Installing Galaxy NG on a machine

Galaxy NG is the upstream version of Automation hub. Its main purpose is to be an on-premises Ansible Galaxy (https://galaxy.ansible.com/) and to be a registry for container images. The GUI is kept in a separate repository as it is not required. It is based on pulp for its services. The latest instructions for installation can be found here: https://github.com/ansible/galaxy_ng/wiki/Development-Setup. The instructions to set up the development environment are as follows:

Clone the repository:

1. Run the following commands to clone the galaxy_ng repository and then change to the newly created directory:

    ```
    $ git clone https://github.com/ansible/galaxy_ng.git
    $ cd galaxy_ng
    ```

2. Use autocomplete on git checkout to find the highest release number:

    ```
    $ git checkout (TAB)
    ```

    The following is an example:

    ```
    $ git checkout v4.2.0
    ```

3. Repeat this for the Galaxy NG UI:

    ```
    $ git clone https://github.com/ansible/ansible-hub-ui.git
    $ cd ansible-hub-ui
    $ git checkout v4.4.1
    ```

Create the compose environment file:

4. Here, the ANSIBLE_HUB_UI_PATH variable is the absolute path to the place you cloned the UI repository. It can be the relative path, but you have to double-check that it is correct compared to what's in the dev/docker-compose.yml file:

    ```
    cat <<EOT >> galaxy_ng/.compose.env
    COMPOSE_PROFILE=standalone
    DEV_SOURCE_PATH='galaxy_ng'
    ANSIBLE_HUB_UI_PATH='/home/user/ansible-hub-ui/'
    EOT
    ```

    Assuming you are in /home/user/, the directory's structure should look like this:

    ```
    ├── galaxy_ng
    │   ├── repository files
    ```

```
|     └── .compose.env
├── ansible-hub-ui
      └── repository files
```

Run the following commands:

1. Change directories to the `galaxy_ng` folder:

    $ cd galaxy_ng

2. Build a Docker image:

    $ ./compose build

3. Start the database and Redis services:

    $ ./compose up -d postgres redis

4. Run migrations:

    $ ./compose run --rm api manage migrate

5. Start the services:

    $ ./compose up -d

    Once everything has run, the instance will be available at `http://localhost:8002`.

This is the base setup for Galaxy NG.

If there is an issue with finding the default password, use the `./compose run --rm api manage createsuperuser` command. This will prompt you to enter the information to create a superuser.

There are a few installation options for Galaxy NG:

- With a GUI, as demonstrated in the previous steps
- Without a GUI
- With a key cloak server
- Building with specific upstream branches

The development link at the beginning of this section provides a rundown of how to customize the installation for those options.

Backing up and restoring the installation

Backup, restore, and secret key redistribution are additional options you have with the installer script. Backup and restore are straightforward since they are ways to back up and restore an existing installation. Always make sure that the installer running the backup or restore matches the version installed. Secret key redistribution is done to update the secret key that's used to encrypt secrets and the database if needed. The backup does a full database backup, while also backing up many relevant files on nodes. With a full backup and a brand new set of machines, it should be possible to fully restore the installation.

Backing up

The backup process requires the inventory file, which works much like the installer. The backup will act on the Automation controller, Automation hub, and the database. In addition to the database, the secret key, custom projects, and configuration files are also held in the backup. A backup should be done before any update is done to a new version.

A backup can be run with the following command:

```
$ ./setup.sh -i inventory -b
```

In addition, there is a role in the controller utilities that will recreate the inventory file and do the necessary backups:

```
- redhat_cop.aap_utilities.backup
```

Using a playbook like this on a scheduled basis will allow you to do regularly scheduled backups. If you wish to only restore Automation hub or the Automation controller alone, then only include the Ansible inventory groups to back up in the inventory file – specifically, the nodes and the database. By omitting the machines from the inventory file, the backup will not know about those nodes, and won't perform a backup. This is useful for backing up the Automation controller and Automation hub separately, to be able to restore them separately as well.

Secret key redistribution

The secret key is important as it is used to encrypt the automation secrets. If it has been compromised, or, as the security policy dictates, it has to be changed periodically, then this can be achieved with the installer as well.

By using the installer with the -k option, a new secret key will be recreated and redistributed to the instances that need it. This inventory file should be the same as the one from the backup. First, back up the instances, and then use the following command to run it:

```
$ ./setup.sh -i inventory -k
```

More information about this can be found here: `https://docs.ansible.com/automation-controller/4.0.0/html/administration/secret_handling.html`.

Restore

The restore works by restoring the data to a freshly installed or already working AAP. This means that the installer needs to be run before the restore can be done. The command to run a restore is as follows:

```
$ ./setup.sh -e 'restore_backup_file=/path/to/nondefault/
location' -r
```

Just like backups, there is a role for doing a restore:

```
- redhat_cop.aap_utilities.restore
```

This role uses the same variables from the installation step to create the inventory file and uses the `restore_location` variable to specify a backup file to restore. It is *not* recommended to automatically kick off the restore, but having a playbook available to run a restore when required is recommended.

Issues that can arise when using the installer

The installers are built to be idempotent so that they can be run multiple times, and always be left with a complete install. However, there will be times when errors occur during installation, leaving things with a partial install. Therefore, when upgrading an installation from one version to another, the best practice is to back up the current installation.

In addition, there are times when it may not seem worth the time to diagnose the problem. So, using VMs in some form allows the user to create snapshots of a clean installation of the operating system before installation.

On every installation, the Ansible playbook output is saved to `setup.log` in the same folder where you ran `./setup.sh`. When an error occurs, you will see an output similar to the following:

cat setup.log

```
TASK [ansible.automation_platform_installer.repo_management :
Enable Automation Platform rhsm repository] ***
fatal: [controller.node]: FAILED! => {"changed": true,
"cmd": ["subscription-manager", "repos", "--enable",
"ansible-automation-platform-2.1-for-rhel-8-x86_64-rpms"],
"delta": "0:17:28.792791", "end": "2022-01-22 22:28:11.
196175", "msg": "non-zero return code", "rc": 70, "start":
"2022-01-22 22:10:42.403384", "stderr": "Network error,
unable to connect to server. Please see /var/log/rhsm/rhsm.log
```

```
for more information.", "stderr_lines": ["Network error, unable
to connect to server. Please see /var/log/rhsm/rhsm.log for
more information."], "stdout": "", "stdout_lines": []}
...ignoring
```

There are two ways to find out which Ansible task failed and, hopefully, find out why it failed.

The first is to use `grep` and find what file the task was in. In this case, `grep -rnw 'Enable Automation Platform rhsm repository'` would give the following response:

```
collections/ansible_collections/ansible/automation_platform_
installer/roles/repo_management/tasks/setup.yml:34:- name:
Enable Automation Platform rhsm repository
setup.log:489:TASK [ansible.automation_platform_installer.repo_
management : Enable Automation Platform rhsm repository] ***
```

By examining the file in the `collection/roles` folder, you can find what the issue is with yum. In this case, there was a connection issue in the network causing the problem, and the installation had to be restarted after that was fixed.

> **Important Tip**
>
> I use this method weekly to find a wide variety of things in Ansible – from tasks to variables, to anything inside a file.

The `grep -rnw 'Enable Automation Platform rhsm repository'` command's options are as follows:

Option	Description
r	Handle directories recursively
n	Print line number with output lines
w	Matches only whole words

Table 2.9 – The grep command's options

By matching the entire task phrase, recursively, and getting the line number, it is fairly easy to find all instances of the search term.

For the most part, the errors in the installation are straightforward when examining the task. With the installer covering so many different services, it is impossible to cover all the issues that may occur. However, the tools in the previous section should prove invaluable in pinpointing any issues that arise.

Summary

In this chapter, you learned how to install many different aspects of AAP. This chapter may seem a bit daunting if you try to do everything. The most important piece to install is the Automation controller.

In the next chapter, you will learn how to install the Ansible Automation controller and Automation hub on Kubernetes or OpenShift. If you are using the virtual machine installation, which we covered in this chapter, then you can skip *Chapter 3, Installing Ansible Automation Platform on OpenShift*. However, it is recommended that you read this and the next chapter to understand the differences between machine and containerized installations.

Once the necessary pieces have been installed, whether you are running a controller or AWX, Automation hub, or Galaxy NG, interacting with them is much the same. If you plan to skip the next chapter, then *Chapter 4, Configuring Settings and Authentication*, goes into the details of setting up authentication and other settings.

3

Installing Ansible Automation Platform on OpenShift

There are a few different ways to install **Ansible Automation Platform** (**AAP**). The two most popular involve installing it on machines, either physically or virtually, or on an OpenShift cluster. In addition, there are upstream options for installing the latest bleeding-edge versions of the Automation controller and Automation hub.

In this chapter, we will cover the following topics:

- Installing AWX on minikube
- Installing CodeReady Containers
- Installing the Red Hat Ansible Automation Platform Operator on an OpenShift cluster
- Backing up and restoring a backup for Automation hub and the Automation controller on OpenShift

Technical requirements

This chapter will cover the platform and methods that will be used in this book. The code referenced in this chapter is available at `https://github.com/PacktPublishing/Demystifying-the-Ansible-Automation-Platform/tree/main/ch03`. It is assumed that you have Ansible installed to run the code provided.

The Ansible Automation Platform Operator requires an OpenShift cluster to be used. While creating and customizing a full OpenShift cluster is beyond the scope of this book, there is an alternative: **CodeReady Containers** (**CRC**). CRC requires a Red Hat Developer Subscription to use. Once you have created CRC, the installation is the same as it would be on OpenShift.

An alternative is to set up an upstream instance of CRC – that is, AWX – on minikube. This chapter will talk about how to create this as well. Although very close to the previous installation method, it is slightly different.

It is highly recommended that you at least follow *one* of the methods from either *Chapter 2, Installing Ansible Automation Platform*, or *Chapter 3, Installing Ansible Automation Platform on OpenShift*, so that you can follow along with the rest of this book.

Installing AWX on minikube

minikube is a local version of Kubernetes that makes it easy to have a development version of Kubernetes on a machine. Kubernetes is an open source system to maintain and scale containerized applications. AWX requires four or more CPUs, 6 GB of free memory, and at least 20 GB of disk space. Due to its small requirements, it can run minikube and AWX on a free GitHub runner instance. The latest information for minikube can be found here: `https://minikube.sigs.k8s.io/docs/start/`. To get minikube running, follow these steps:

1. Run the following commands to download the installation file, install minikube, and get it running:

```
$ curl -LO https://storage.googleapis.com/minikube/
releases/latest/minikube-linux-amd64
$ sudo install minikube-linux-amd64 /usr/local/bin/
minikube
$ minikube start --cpus=4 --memory=6g --addons=ingress
```

2. Once minikube is running, make sure it is working properly:

```
$ minikube kubectl -- get nodes
$ minikube kubectl -- get pods -A
```

3. Use the minikube dashboard to get a visual representation of the installation:

```
$ minikube dashboard
```

4. To make things easier, make an alias of the `kubectl` command:

```
$ alias kubectl="minikube kubectl --"
```

Once minikube is ready, it is time to install AWX. This can be done with `awx-operator`. The instructions for installation can be found here: `https://github.com/ansible/awx-operator#basic-install`. It involves cloning either the development branch or a release branch of the operator. Follow one of the following options to clone the correct repository branch. If you have problems with the development branch, switch to a release branch.

Clone the release branch:

1. Navigate to `https://github.com/ansible/awx-operator/releases`.

2. Click on the latest release number, which should lead to a link similar to `https://github.com/ansible/awx-operator/releases/tag/0.15.0`.

3. Download the tarball, untar it, and change directories:

    ```
    $ tar xvf awx-operator-0.15.0.tar.gz
    $ cd awx-operator-0.15.0.tar
    ```

Clone a branch based on the release label:

4. Run the following commands:

    ```
    $ git clone https://github.com/ansible/awx-operator.git
    $ cd awx-operator
    ```

5. Use autocomplete on `git checkout` to find the highest release number:

    ```
    $ git checkout (TAB)
    ```

    For example, you could use the following command:

    ```
    $ git checkout 0.15.0
    ```

Clone a branch based on the development branch:

6. Run the following commands:

    ```
    $ git clone https://github.com/ansible/awx-operator.git
    $ cd awx-operator
    ```

Now that the repository has been cloned, it is time to set up the operator:

7. Export the namespace, as follows:

    ```
    $ export NAMESPACE=my-namespace
    ```

8. Use the `make deploy` command to build the deployment:

    ```
    $ make deploy
    ```

    Wait a while and ensure that it has been deployed correctly.

9. To check the status of the deployment, run the following `kubectl` command. This will provide the status of the pods that have been created for the deployment:

    ```
    kubectl get pods -n $NAMESPACE
    ```

10. To make things easier, set a default namespace:

```
kubectl config set-context --current
--namespace=$NAMESPACE
```

11. Using the awx-demo.yml file in the repository, create a file of your own or use the awx-demo.
yml file in this chapter's resources. metadata.name will be the name of your AWX instance.
The following code shows an example of this:

```
---
apiVersion: awx.ansible.com/v1beta1
kind: AWX
metadata:
    name: awx-demo
spec:
    service_type: nodeport
```

12. Use kubectl to apply the configuration file:

```
$ kubectl apply -f awx-demo.yml -n $NAMESPACE
```

13. You can monitor the deployment by running various commands, as follows:

- Use the following command to check the logs:

```
$ kubectl logs -f deployments/awx-operator-controller-
manager -c awx-manager
```

- Use the following command to check the pods:

```
$ kubectl get pods -l "app.kubernetes.io/managed-by=awx-
operator"
```

- Use the following command to check the services:

```
$ kubectl get svc -l "app.kubernetes.io/managed-by=awx-
operator"
```

14. After a little while, the deployment should be complete. Check it by using the following command:

```
$ minikube service awx-demo-service --url -n $NAMESPACE.
```

This should return a URL such as http://192.168.39.74:30261.

15. To get the admin password, use the following command:

```
$ kubectl get secret awx-demo-admin-password -o
jsonpath="{.data.password}" | base64 -decode
```

Remember, if you changed `metadata.name` in your deployment file, all references to `awx-demo` will need to be changed as well for these commands to work.

16. Navigate to the URL from *step 8*, and the password from *step 9*, and log in with a browser using the *admin* account.

A working installation of AWX allows a local development environment to be created. While it is not 1:1 in terms of CRC, it is extremely similar. It is very useful for development in a home lab.

Now that we have learned how to install AWX on minkube, let's look at using CodeReady Containers. This is an alternative to minikube and is an OpenShift deployment that allows you to install Ansible Automation Platform.

Installing CodeReady Containers

CodeReady Containers is a single-node OpenShift cluster designed for use on a laptop. The latest way to install can be found at `https://developers.redhat.com/products/codeready-containers/getting-started`. If the instructions here do not work, please refer to the documentation. Documentation is available for installation on Linux, Windows, or Mac.

To install CRC on an RHEL-based Linux machine, follow these steps:

1. Make sure that you have `NetworkManager` installed by running the following command:

```
$ sudo dnf install NetworkManager
```

2. Navigate to `https://console.redhat.com/openshift/create/local` and download the latest release.

3. Navigate to where you downloaded the latest release and run the following command:

```
$ tar xvf crc-linux-amd64.tar.xz
```

4. Create a `bin` folder, copy the `crc` file into the bin folder, and then add the `bin` path to your path:

```
$ mkdir -p ~/bin
$ cp crc ~/bin
$ export PATH=$PATH:$HOME/bin
$ echo 'export PATH=$PATH:$HOME/bin' >> ~/.bashrc
```

5. Configure how much memory or CPU you want to give your CRC instance and run the setup:

```
$ crc config set cpus 5
$ crc config set memory 9126
$ crc setup
$ crc start
```

Now, the CRC instance should be available at `https://console-openshift-console.apps-crc.testing`. Use the information from the `crc start` command to access the console. If you forget the login password, it can be retrieved using the `crc console -credentials` command. The next step is to follow the OpenShift installation instructions in the next section to install the Ansible Automation Platform Operator.

Installing the Red Hat Ansible Automation Platform operator on an OpenShift cluster

Using an OpenShift cluster to host Automation hub and the Automation controller provides benefits for scaling and integrates well with your existing OpenShift cluster. With either OpenShift or CRC, the process is the same.

Follow these steps to install the Ansible Automation Platform Operator:

1. Navigate to the web console of the OpenShift instance and log in.

2. In the web GUI, navigate to **Operators** | **Operator Hub**.

3. In the keyword text box, enter `Ansible`.

4. Click on **Ansible Automation Platform** and select **Install**.

5. Choose one option from the following selections:

    A. Update channel: Which release to use. Generally, you should use the latest version:

    • stable-2.2

    • stable-2.1

    • Early Access – v2.0.1

    B. Scope:

    • Namespace scoped (default): Watches and manages resources in a single namespace

    • Cluster scoped: Watches and manages resources cluster-wide

    C.   These combinations of release and scope will be combined to create the following options to choose from:

- **early-access**
- **early-access-cluster-scoped**
- **stable-2.1**
- **stable-2.1-cluster-scoped**
- **stable-2.2**
- **stable-2.2-cluster-scoped**

    D.   Select either a default namespace or a project of your choice.

    E.   Choose **Automatic** or **Manual** updates for the Operator. I recommend **Automatic** as this will only update the operator.

6.   Click **Install**. After a few moments, navigate to **Installed Operators**, and then to **Ansible Automation Platform**.

The operator manages Automation controller and Automation hub deployments. One key difference with Automation hub is that no container registry is created. It is assumed that if OpenShift is being used, a registry already exists.

With a CRC deployment created, the next step is to add the Automation controller or hub to it.

Exploring the Automation controller and Automation hub on OpenShift

The controller can be installed through the web interface through the GUI or YAML, or a definition file in the CLI. For both the Automation controller and Automation hub, the operator GUI gives you a form to fill out. The variables are presented in a flattened view, in segments. This chapter's repository (`ch03/operator_files/`) contains additional references to be used for deployment.

The resource containers have limits and requests. More details on what the values mean and how they apply can be found here: `https://docs.openshift.com/container-platform/3.11/dev_guide/compute_resources.html#dev-memory-requests`.

The following tables are by no means a comprehensive list of all variables used in the Ansible Automation Platform Operator, a more comprehensive guide to the variables can be found here: `https://github.com/ansible/awx-operator#service-type`.

First, there are the resource variables. These all have entries regarding limits and requests to determine the size of the containers they operate with.

The ones for the Automation controller are as follows:

Variable	Description
ee_resource_requirements	Resource requirements for the ee container
task_resource_requirements	Resource requirements for the task container
web_resource_requirements	Resource requirements for the web container
postgres_storage_requirements	Storage requirements for the PostgreSQL container
postgres_resource_requirements	Resource requirements for the PostgreSQL container

Table 3.1 – Automation controller Operator container variables

The ones for Automation hub are as follows:

Variable	Description
resource_manager	The pulp resource manager deployment
redis_resource_requirements	Resource requirements for the Redis container
worker	The pulp worker deployment
api	The pulp api deployment
web	The pulp web deployment
content	The pulp content deployment
postgres_storage_requirements	Storage requirements for the PostgreSQL container
postgres_resource_requirements	Resource requirements for the PostgreSQL container

Table 3.2 – Automation hub Operator container variables

Each of these contains sub-variables called *requests* and *limits*, which are the respective constraints on the containers: what is requested and what it is limited to. A good breakdown of their usage in OpenShift can be found at https://docs.openshift.com/container-platform/4.6/nodes/clusters/nodes-cluster-limit-ranges.html.

Each of these sub-variables uses variables to dictate CPU, memory, and storage. They are as follows:

Variable	Example	Description
cpu	250m	CPU resources to use
memory	50Mi	Bytes with SI notation
ephemeral-storage	50Mi	Bytes with SI notation
replicas	1	Number of containers to run
log_level	INFO	Log level to use

Table 3.3 – Automation controller Operator container resource variables

These are organized in YAML format for each resource. An example of a hub web resource is as follows:

operator_files/automation_hub_all.yml

```
web:
  log_level: INFO
  replicas: 1
  resource_requirements:
    limits:
      cpu: 500m
      memory: 50Mi
      ephemeral-storage: 50Mi
    requests:
      cpu: 500m
      memory: 50Mi
      ephemeral-storage: 50Mi
```

More examples such as this can be found in the chapter's code folder.

Learning about Automation controller specifics on OpenShift

All these variables are either set in the GUI or a YAML definition file. This will create a controller. Each can be grouped into a specific category.

Secret fields

Some variables point to OpenShift secrets. Secrets are objects in OpenShift that hold sensitive data. These variable fields point at OpenShift objects that contain the actual value that the Operator uses:

Secret fields	Notes
admin_password_secret	Admin password
image_pull_secret	Secret to pull images from the registry
postgres_configuration_secret	Database password secret
old_postgres_configuration_secret	If migrating from an old database, postgres password
secret_key_secret	If preset, the secret key secret
bundle_cacert_secret	Secret that has the trusted Certificate authority bundle
ee_pull_credentials_secret	Secret to pull images for execution enviroments from registery
ldap_cacert_secret	Secret that has the ldap certification information

Table 3.4 – Ansible Operator controller secret variables

These secret fields are used to access sensitive information, the next set of fields are the base variables.

Base variables

The following table shows the base variables for the controller:

Variable Name	Default Value	Description
name	demo	Name of deployment
namespace		Namespace to put deployment in
create_preload_data	TRUE	Preload basic data such as demo project and job template
admin_user	admin	Admin username
hostname		FQDN hostname of the controller when using ingress
replicas	1	Number of controller control nodes to deploy
garbage_collect_secrets	FALSE	Whether or not to remove secrets upon instance removal

Table 3.5 – Ansible Operator controller base variables

After the base variables, the next set to examine are those that determine ports used.

Port variables

The following table shows the variables that address the ports to use for the operator:

Variable Name	Default Value	Description
loadbalancer_protocol	http	Loadbalancer protocol
loadbalancer_port	80	Loadbalancer port
nodeport_port	30080	Port to use for nodeport

Table 3.6 – Ansible Operator controller port variables

After port variables, the next set are project variables where the Automation controller stores projects.

Project variables

The following table shows the variables for the storage class and other project-based variables:

Variable Name	Default Value	Description
projects_storage_class		Storage class to use for projects
projects_storage_size	8Gi	Storage size for projects
projects_storage_access_mode	ReadWriteMany	Projects storage access mode
projects_persistence	TRUE	Whether project data is stored in persistent memory

Table 3.7 – Ansible Operator controller project variables

With project storage squared away, there are a few more variables used that do not fit neatly into a category.

Miscellaneous variables

The following table shows all other controller variables:

Variable Name	Default Value	Options	Description
service_type	NodePort	NodePort |Cluster IP |Loadbalancer	How traffic gets into the cluster of services
ingress_type	none	None | Ingress | Route	The ingress type to use to reach the deployed instance
image_pull_policy	IfNotPresent	Always | Never | IfNotPresent	Pull policy for images
route_tls_termination_mechanism	Edge	Edge | Passthrough	TLS termination mechanism
task_privileged	FALSE		Whether tasks pods are privileged
admin_email			Email address of the admin
service_account_annotations			Annotations for the service account in OpenShift

Table 3.8 – Ansible Operator controller misc. variables

These variables are used to define the Automation controller deployment, the next section will go into detail on how to use them.

Defining the Automation controller

The following is an excerpt from the Kubernetes definition file. It uses variables from the aforementioned definitions. A full example of the Operator definition can be found in this chapter's GitHub repository:

ch03/controller_base.yml

```
---
apiVersion: automationcontroller.ansible.com/v1beta1
kind: AutomationController
metadata:
  name: crc-controller
  namespace: ansible-automation-platform
spec:
  create_preload_data: true
  route_tls_termination_mechanism: Edge
  loadbalancer_port: 80
  projects_storage_size: 8Gi
  task_privileged: false
  replicas: 1
  admin_user: admin
  admin_password_secret: builder-dockercfg-tbbzq
...
```

Once the operation is complete in a basic installation, a total of five containers will be created – one Postgres, one Redis, one web, one task, and one execution environment. To get to the web address of the controller, navigate to **Networking | Routes**. Next to the deployment name of the controller, you should be able to find the URL's location. For a CRC installation and the aforementioned deployment, the URL would be `https://crc-controller-ansible-automation-platform. apps-crc.testing/`. To find the admin secret that houses the password to log in, navigate to **Workloads | Secrets** and look for `crc-controller-admin-password`. At the bottom of the **Secrets** page, the password can be revealed.

Learning Automation hub specifics

While both Automation hub and the Automation controller use the same variables to describe their containers, things shift with the other variables. All these variables are either set in the GUI or a YAML definition file. This will create an Automation hub. Each can be grouped into a specific category.

Secret fields

Secrets are where passwords are stored on OpenShift. The following table shows these secrets:

Secret Fields	Description
sso_secret	Configuration secret for the SSO instance
admin_password_secret	Admin password
postgres_migrant_configuration_secret	Secret where the old database configuration can be found for data migration
postgres_configuration_secret	Secret where the database configuration can be found
db_fields_encryption_secret	Secret where the Fernet symmetric encryption key is stored
image_pull_secret	Image pull secret for container images

Table 3.9 – Ansible Operator hub secret variables

These variables are used to access sensitive information. The next set are the base variables to use.

Base variables

The following table shows the base variables for Automation hub:

Name	Example	Description
namespace		Namespace to put the deployment in
labels		Labels to apply to Automation hub, Open Shift standard
hostname	hostname	The hostname of the instance
image_pull_policy	IfNotPresent	Pull policy for images

Table 3.10 – Ansible Operator hub base variables

With the base variables defined, the next section involves storage.

Storage variables

The following table shows the variable addresses that the storage options use for the Operator:

Name	Example	Description
file_storage_size	100Gi	The size of the file storage; for example 100 Gi
file_storage_access_mode	ReadWriteMany	The file's storage access mode
storage_type	File	Configuration for the storage type to be utilized
file_storage_storage_class		Storage class to use for the file persistentVolumeClaim

Table 3.11 – Ansible Operator hub storage variables

With storage taken care of, the next set of variables involves which ports to use to access the Automation hub.

Port variables

The following table shows the variable addresses that the ports use for the Operator:

Name	Example	Description
route_host		The DNS to use to point to the instance
route_tls_termination_mechanism	Edge | Passthrough	The secure TLS termination mechanism to use
ingress_type	None | Ingress | Route | NodePort | Loadbalancer	The ingress type to use to reach the deployed instance
loadbalancer_protocol	http	Protocol to use for the load balancer
loadbalancer_port	80	Port to use for the load balancer
nodeport_port	80	Provide requested port value
loadbalancer_annotations		Annotations to add to the load balancer
ingress_annotations		Annotations to add to the ingress

Table 3.12 – Ansible Operator hub port variables

These variables are all used in the Automation hub deployment, the next section will go into detail on how to use them.

Defining Automation hub

The following is an excerpt from the Kubernetes definition file for Automation hub. It uses variables from the definitions provided in *Tables 3.9* to *3.12*. A full example of the Operator definition can be found in this chapter's GitHub repository:

ch03/automation_hub_base.yml

```
---
apiVersion: automationhub.ansible.com/v1beta1
kind: AutomationHub
metadata:
 name: crc-hub
 namespace: ansible-automation-platform
spec:
 route_tls_termination_mechanism: Edge
 loadbalancer_port: 80
 file_storage_size: 100Gi
 image_pull_policy: IfNotPresent
 web:
   replicas: 1
 file_storage_access_mode: ReadWriteMany
 api:
   log_level: INFO
   replicas: 1
 loadbalancer_protocol: http
...
```

In addition to the seven containers mentioned previously (one API, two content, one resource manager, one web, and two workers), the operator also created a Redis and a Postgres container. To get to the web address of Automation hub, navigate to **Networking** | **Routes** and find the URL of Automation hub next to the deployment name. For example, when using a CRC, the web address for what we did here would be `https://crc-hub-ansible-automation-platform.apps-crc.testing/`. The admin secret that houses the password can be found by navigating to **Workloads** | **Secrets** and looking for `crc-hub-admin-password`. At the bottom of the secret page, the password can be revealed.

Backing up and restoring a backup for Automation hub and the Automation controller on OpenShift

You can back up Automation hub or the Automation controller with the GUI or a YAML definition. Here, you must provide the name of the backup and the deployment name to back up, as follows:

```
---
apiVersion: automationhub.ansible.com/v1beta1
kind: AutomationHubBackup
metadata:
 name: example
 namespace: ansible-automation-platform
spec:
 deployment_name: crc-hub
...
```

The preceding variables, along with other spec variables, are described in the following table:

Variable	Description
backup_pvc	Name of the PVC to be used for storing the backup
backup_pvc_namespace	Namespace that the PVC is in
backup_storage_class	Storage class to use when creating a PVC for backup
backup_storage_requirements	Storage requirements for the backup
deployment_name	Name of the deployment to be backed up
Name	Name of the backup
kind	[AutomationControllerBackup]	[AutomationHubBackup] - what type of backup to use

Table 3.13 – Ansible Operator backup variables

These allow the user to determine how much and what type of space to use, and where to store the backup.

restore is done similarly:

```
---
apiVersion: automationhub.ansible.com/v1beta1
kind: AutomationHubRestore
metadata:
 namespace: ansible-automation-platform
 name: restore
spec:
 backup_name: backup
```

```
backup_source: CR
. . .
```

This `restore` definition relies on the previously created custom resource to point at the backup. It can also be done by supplying variables for a PVC. All these variables can be found in the following table:

Variable	Description
backup_dir	Backup directory name, set as a status found on the backup object (backupDirectory)
backup_name	Backup name
backup_pvc	Name of the PVC to be restored from, set as a status found on the backup object (backupClaim)
backup_pvc_namespace	Namespace the PVC is in
backup_source	PVC | CR; source either persistent volume or custom resource
deployment_name	Name of the deployment to be restored to
storage_type	Configuration for the storage type utilized in the backup
Name	Name of the backup
kind	[AutomationControllerRestore] | [AutomationHubRestore] - what type of backup was used.

Table 3.14 – Ansible Operator restore variables

The backup directory and other variable values can be found in the resource of the backup that was previously completed.

Summary

This chapter explained how to install the many different aspects of Ansible Automation Platform on some form of Kubernetes. This chapter may seem a bit daunting if you try to do everything at once. The most important piece to install is CRC, which you can do by following either this chapter's or the previous chapter's instructions.

Once the pieces have been installed, whether you are running CRC or AWX, Automation hub, or Galaxy NG, interacting with them is similar. The next chapter will explore the basic setup post-installation.

4

Configuring Settings and Authentication

Once an Automation controller and hub are installed, they need to be configured for authentication. This chapter will focus on authentication and the creation of users and groups. The Automation controller itself allows for the creation of users, teams, and organizations for basic purposes, but for an enterprise with a large number of users, tying it into an external system such as **Lightweight Directory Access Protocol** (**LDAP**) is much more flexible, and easier to scale. In addition, there are settings within the Automation controller that can be set that modify its behavior.

This chapter will cover two main subjects. The first is configuring Automation controller settings. This section will cover how to register the Automation controller with a subscription manifest, and the process of updating Automation controller settings. The second section will cover the authentication configuration of the Automation controller and hub using the **Red Hat Single Sign-On** (**RH-SSO**) server. This includes getting values from a Windows **Active Directory** (**AD**) server, configuring the SSO server for **Security Assertion Markup Language** (**SAML**), setting up Automation hub administrators with SSO, and integrating the Automation controller with SSO. The SSO server is the focus for authentication as it is the central place to allow authentication for the whole Automation Platform.

We will cover the following topics in this chapter:

- Configuring Automation controller settings
- Configuring the RH-SSO server SAML
- Integrating LDAP with Microsoft AD
- Adding users and teams to the Automation controller without an **identity provider** (**IdP**)
- Adding users and groups to the Automation hub without an IdP

Technical requirements

This chapter will go over the platform and methods used in this book. All code referenced in this chapter is available at `https://github.com/PacktPublishing/Demystifying-the-Ansible-Automation-Platform/tree/main/ch04`.

It is assumed that you have Ansible installed in order to run the code provided. In addition, it is assumed that both an Automation controller and hub have been installed, This chapter will also cover pairing the SSO server with the hub and controller, though that part is optional.

Getting a trial version of Ansible Automation Platform

A trial version of Ansible Automation Platform can be obtained with a Red Hat developer account. To get a developer account, register at `https://developers.redhat.com/`. After registering for a developer account, navigate to `https://www.redhat.com/en/technologies/management/ansible/try-it`. This will populate a subscription at `https://access.redhat.com/management/subscriptions`. This will be needed post installation when first setting up an Automation controller.

Configuring Automation controller settings

Automation controller settings are where the access controls are stored in the controller. However, before a controller can be accessed, a subscription must be added. This section will cover how to go through registration and change settings inside of the controller. The section after that will go into detail about authentication settings.

Registering the Automation controller with a subscription manifest

When an Automation controller is first accessed, it asks the user to activate a subscription. This can be done with either a manifest file from the Red Hat website, from a Satellite installation, or using the user's Red Hat username and password. The registration page looks like this:

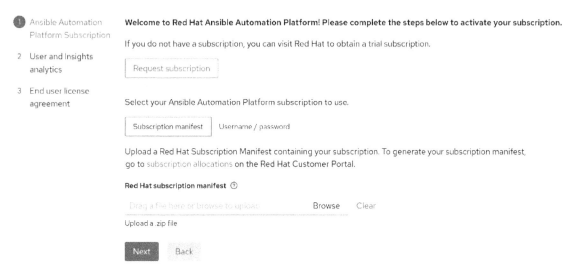

Figure 4.1 – Automation controller registration

Using a username and password for this process will present the user with a selection of subscriptions to choose from. However, if a subscription is being split between multiple servers, or there is a network issue between the Ansible controller and Red Hat servers or Satellite servers, then a manifest file is a way to provide a subscription. Instructions on obtaining a manifest file can be found at https://docs.ansible.com/automation-controller/latest/html/userguide/import_license.html#obtain-sub-manifest.

This manifest file can then be uploaded to the web **graphical user interface** (**GUI**) of the Automation controller. It's also possible to use modules and roles to push the manifest to the controller. We'll now look at some examples of how to do this.

Here's how to update the manifest using the ansible.controller module:

```
// license/aap_license_module.yml
---
- name: Update manifest to controller
  hosts: localhost
  connection: local
  gather_facts: false
  collections:
    - ansible.controller
  tasks:
    - name: Push Manifest
      ansible.controller.license:
```

```
manifest: manifest_file.zip
force: true
controller_host: https://controller.node
controller_username: admin
controller_password: secret123
validate_certs: false
```

Using this module is good for running one-off updates of the license to the Automation controller. However, with the goal of **configuration as code** (**CaC**), it is usually preferable to create a variable file with the license location that is invoked, which can be consumed by the `redhat_cop.controller_configuration` role. The use of the configuration file is illustrated here:

```
settings/config/license.yml
---
    controller_license:
      manifest_file: manifest_file.zip
      force: true
```

The role can then be invoked using this code:

```
//settings/aap_settings_role.yml
  pre_tasks:
    - name: Include vars from configs directory
      include_vars:
        dir: ./configs  roles:
    - redhat_cop.controller_configuration.license
```

These methods can be used to create Automation controller instances using scripts and pipelines.

Updating Automation controller settings

Automation controller settings dictate how authentication is handled and provide global options for jobs and logging, along with various system settings. As of the time of writing, including all authorization fields, there are around 250 settings that can be set. Generally, these do not need to be changed, but there are circumstances in which this is required.

For example, if using an F5 load balancer, the remote host header setting needs to be updated to this:

```
REMOTE_HOST_HEADERS = ['HTTP_X_FORWARDED_FOR', 'REMOTE_ADDR',
'REMOTE_HOST']
```

Another two popular settings to change are shown here:

```
"CUSTOM_LOGIN_INFO": "",
"CUSTOM_LOGO": "",
```

These two settings will change the text and logo used on the landing page of the Automation controller. The text is commonly changed to include a friendly greeting or legal warning about unauthorized use. For note, the logo would point to an image hosted somewhere the controller can resolve.

The next sections will focus on how to update settings for the Automation controller using modules and a predefined role.

Changing Automation controller settings using controller modules

To make a one-off change to a task in a playbook for settings, it is best to use a module. Here's how to do this using the ansible.controller module:

```
// settings/aap_settings_module.yml
---
  - name: Change settings
    ansible.controller.settings:
      settings:
        SESSION_COOKIE_AGE: 3600
        CUSTOM_LOGIN_INFO: "Welcome to the Automation
Controller"
```

Changing Automation controller settings using redhat_cop roles

Predefined configuration files are the ideal way to use a role. The configuration file for settings has the following format:

```
// settings/config/settings.yml
---
controller_settings:
  settings:
    SESSION_COOKIE_AGE: 3600
    CUSTOM_LOGIN_INFO: "Welcome to the Automation Controller"
...
```

To invoke a role in a playbook and include the files in the configuration chapter, use the following Ansible tasks:

```
// settings/aap_settings_role.yml
  pre_tasks:
    - name: Include vars from configs directory
      include_vars:
        dir: ./configs
        ignore_files: [controller_config.yml.template]
        extensions: ["yml"]
      tags:
        - always
  roles:
    - redhat_cop.controller_configuration.license
    - redhat_cop.controller_configuration.settings
```

It is assumed that in later portions of this chapter, one of these three methods will be used to push the settings to the controller for configuring authentication.

> **Important Note**
> ALLOW_OAUTH2_FOR_EXTERNAL_USERS may need to be set to true for some users accounts to be used for authentication in redhat_cop and ansible.controller roles and modules if using external authentication methods later in this chapter.

The most important settings for the Automation controller are those used to configure authentication. The rest of the chapter will go over different methods for how to do this.

Configuring the RH-SSO server SAML

This section will cover the configuration of adding an LDAP server to the RH-SSO server and then linking the SSO server to the Automation controller. As of Ansible Automation Platform 2.1.1, the installer will automatically install, configure, and integrate the SSO server with the Automation hub, but not the Automation controller. An alternative to setting up the initial realm can be found here: https://www.ansible.com/blog/red-hat-single-sign-on-integration-with-ansible-tower. However, it is recommended to use the installer.

Getting values from a Windows AD server

Many users have Windows AD systems that are tied into Ansible Automation Platform. Some common values need to be obtained in order to integrate the two systems. First is the LDAP server itself, which is the domain server. Normally, this is how you would obtain the server's name:

```
Ldap://Servername
```

But if that is not working, use the Windows search bar, type `ADSIedit` to open that program, click **Action | Connect to…**, and the server path should be displayed.

The next piece needed is the `Users` and `Bind` **distinguished name** (**DN**). Most LDAP systems do not allow for anonymous access, so the bind DN is the DN to authenticate as. Normally, these are similar as they are both user accounts. Because an administrator account is needed to authenticate for the bind DN, the **command-line interface** (**CLI**) command to use on the Windows server is `dsquery user -name Administrator`.

That should output the bind DN of all users in the LDAP tree, as follows:

```
CN=Administrator,CN=Users,DC=ansible,DC=lcl
```

Taking out the `CN=Administrator` portion gives us the `Users` DN, as shown here:

```
CN=Users,DC=ansible,DC=lcl
```

Finally, there is the **organizational unit** (**OU**) that is used to import user groups, which is the OU in the domain component, as shown here:

```
OU=Ansible,DC=ansible,DC=lcl
```

These values are needed later to connect either the Ansible controller or the RH-SSO server to the AD server.

Configuring the RH-SSO server

The RH-SSO server is an option to use with the Automation controller. It is the only method to integrate LDAP and other outside authentication services with the Automation hub. It allows for authentication integration with SAML, LDAP, GitLab, Google, Microsoft, and a host of other services. More information can be found at `https://access.redhat.com/documentation/en-us/red_hat_single_sign-on/`.

To connect an LDAP server to an SSO server, the following steps are taken:

1. Navigate to the SSO server web page. This is normally found at `https://servername:8443/`.

2. Log in with the username and password used during installation, which is a username of `admin` and the variable value from the installer of `sso_console_admin_password`.

3. Navigate to the **Administration** console | **User Federation** | **Add provider** | **LDAP**.

4. Set the following options. Most of these do not need to be changed, but these are the ones that need to be set for AD:

Setting	Value	Description
Console Display Name	ldap	Display name of provider
Vendor	Active Directory	Vendor to use Red Hat or Active Directory
Connection URL	LDAP://servername	Ldap server connection
Users DN	CN=Users,DC=ansible,DC=lcl	Full DN of the Ldap Tree where the users are
Bind DN	CN=Administrator,CN=Users,DC=ansible,DC=lcl	DN of Ldap Admin
Bind Credential	password	The password for the BIND DN user

Figure 4.2 – User federation settings

This is how it should look in the form:

Figure 4.3 – SSO LDAP settings

5.   Click the **Test Connection** and **Test Authentication** buttons to verify that the SSO server can both connect and authenticate to the LDAP server.

6.   Click **Save** to finalize the configuration.

To map groups to roles, continue from the previous screen and do the following:

1.   Navigate to **Mappers | Create**.

2.   Name the mapper, and choose the `role-ldap-mapper` option.

3.   In the **LDAP Roles DN** field, insert the OU to map for groups—for example, `OU=Ansible,DC=ansible,DC=lcl`.

4.   Click **Save**.

5.   It should look like this in the form:

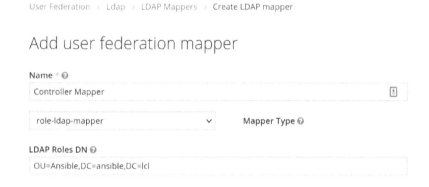

Figure 4.4 – User federation mapping

Now that user federations have been created, a client profile needs to be created for the controller. This includes mapping values the RH-SSO server has imported from LDAP and mapping them to specific values for the Automation controller. Here's how to do this:

1.   Navigate to **Clients | Create**.

2.   Fill in a client **identifier (ID)** with a unique name to identify the client, preferably something to do with the controller.

3.   Select **Client Protocol** as **SAML**, then click **Save**.

4.   Fill in the form with the values shown here:

Fields	Value	Description
Client ID	`controller.node`	Unique name for the client
Root URL	`https://controller.node/`	URL of the controller
Valid Redirect URIs	`https://controller.node/sso/complete/saml/`	The redirect URL from controller
IDP Initiated SSO URL Name	`controller.node`	FQDN of the controller
Fine Grain SAML Endpoint Configuration		Advanced option to expand
Assertion Consumer Service POST Binding URL	`https://controller.node/sso/complete/saml/`	Binding URI to send response to

Figure 4.5 – Client creation variables

5. When completed, the fields in the form should look like this (this is a reduced view of the screen, and unused fields have been omitted):

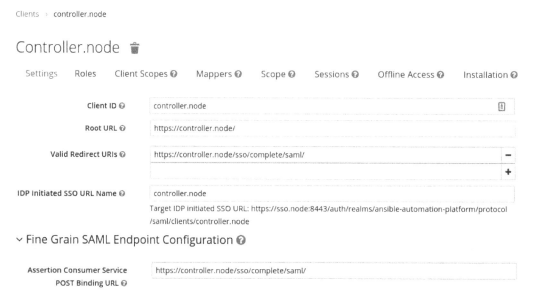

Figure 4.6 – SSO client settings

6. Click **Save**.

7. Navigate to **Mappers**.

8. Create a mapper for each of the following configurations: user_name, last_name, email, user_permanentID, and first_name, as shown in the following table. Click **Save** after creating each mapped entry:

Name	Mapper Type	Property	Friendly Name	SAML Attribute Name	SAML Attribute NameFormat
email	User Property	email	Email	email	Basic
last_name	User Property	lastName	Last Name	last_name	Basic
first_name	User Property	firstName	First Name	first_name	Basic
username	User Property	username	User Name	username	Basic
user_permanent_id	User Property	uid	name_id	name_id	Basic

Figure 4.7 – Client mapper variables

> **Important Note**
>
> firstName and lastName are case-sensitive, and they map to the RH-SSO user property.

9. In addition, the following need to be set as additional mappers, with slightly different fields:

Name lame	Mapper Type	Group Attribute Name	Friendly Name	SAML Attribute NameFormat	Single Group Attribute	Full Group Path
group	Group list	group	group	Basic	On	Off
role list	Role list	Role	Role	Basic	On	Off

Figure 4.8 – Client mapper group and role variables

All of these fields map values from SAML to the Automation controller. This concludes the configuration of the SSO server.

Setting Automation hub administrators with RH-SSO

Setting administrators for the Automation hub in the RH-SSO server is done through role mapping. To give a user or group this role mapping, follow the next steps:

1. Navigate to the `ansible-automation-platform` realm in the SSO client.

2. Navigate to the **Manage** section on the left-hand side and choose **Users** or **Groups**.

3. Click either the ID of the user or, if using a group, click the name of the group, and click **Edit**.

4. Navigate to **Role Mappings**.

5. Using the drop-down menu of **Client Roles**, select **automation-hub**.

6. Click **hubadmin** from the **Available Roles** field, then click **Add selected**.

This is how it should look in the form. Once again, some fields have been omitted:

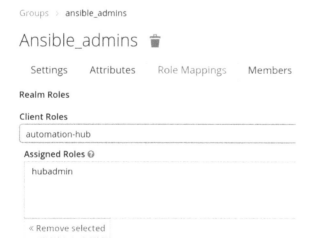

Figure 4.9 – SSO role mapping

This concludes configuring the SSO server. The next step is to configure the Automation controller to interact with the SSO server. These configurations need to be done on both services to complete the process.

Integrating the Automation controller with RH-SSO

Now that the RH-SSO server is configured, it can now act as an IdP of the Automation controller. During the installation process, a **Rivest-Shamir-Adleman** (**RSA**) key and certificate were created. These will need to be retrieved for use.

In order to get the key and certificate, proceed as follows:

1. On the SSO server, navigate to **Realm Settings** | **Keys**.

2. Click on **Public key** and **Certificate** and save the values.

On the Automation controller in the GUI at **Settings** | **SAML Settings**, there are values to be set. This can also be done via the **application programming interface** (**API**) modules and roles, as discussed earlier in this chapter in the *Configuring Automation controller settings* and *Updating Automation controller settings* sections. The following table shows how to translate GUI fields to their API values:

GUI Field	API Setting
SAML Service Provider Entity ID	SOCIAL_AUTH_SAML_SP_ENTITY_ID
SAML Service Provider Public Certificate	SOCIAL_AUTH_SAML_SP_PUBLIC_CERT
SAML Service Provider Private Key	SOCIAL_AUTH_SAML_SP_PRIVATE_KEY
SAML Service Provider Organization Info	SOCIAL_AUTH_SAML_ORG_INFO
SAML Service Provider Technical Contact	SOCIAL_AUTH_SAML_TECHNICAL_CONTACT
SAML Service Provider Support Contact	SOCIAL_AUTH_SAML_SUPPORT_CONTACT
SAML Enabled Identity Providers	SOCIAL_AUTH_SAML_ENABLED_IDPS

Figure 4.10 – GUI to API field equivalents

The variables set for the Automation controller should be self-explanatory. However, here are some details about the upcoming variables:

*   The **entity ID** must much that set in the client ID previously set.

*   RH-SSO is a visual marker and is not tied to the **Domain Name System** (**DNS**) or a configuration.

*   The **Technical Contacts** information must be filled out, but this can be any valid name and email.

*   The **entity ID** and **Uniform Resource Locator** (**URL**) from IdPs must be the realm created earlier. `ansible-automation-platform` is the realm created with the installer.

*   The `attr_*` value should match the mapped values previously set.

- For these three values, use the values for the certificate and key from the beginning of this section:

  - `SOCIAL_AUTH_SAML_SP_PUBLIC_CERT`

  - `SOCIAL_AUTH_SAML_SP_PRIVATE_KEY`

  - `x509cert`

  Insert the certificate/key where it says **Key/Cert Value**.

> **Important Note**
>
> The `x509cert` variable just shown is a string and must be kept as a string. Using multiline with newline characters or a list of strings breaks things in the API as it expects a string.

The settings for integrating with the RH-SSO server are shown in the following file:

```
// sso/saml/settings.yml
---
controller_settings:
  settings:
```

The following code sets the certificates and authorization to the SSO server:

```
SOCIAL_AUTH_SAML_SP_ENTITY_ID: controller.node
SOCIAL_AUTH_SAML_SP_PUBLIC_CERT:
SOCIAL_AUTH_SAML_SP_PRIVATE_KEY:
```

In addition, this code links to an account:

```
SOCIAL_AUTH_SAML_ORG_INFO:
  en-US:
    displayname: RHSSO
    url: https://sso.node:8443
    name: rsa-generated
SOCIAL_AUTH_SAML_TECHNICAL_CONTACT:
  givenName: Person
  emailAddress: person@org.com
SOCIAL_AUTH_SAML_SUPPORT_CONTACT:
  givenName: Person
  emailAddress: person@org.com
```

These are the settings for the SAML IdP:

```
SOCIAL_AUTH_SAML_ENABLED_IDPS:
  RHSSO:
    entity_id: https://sso.node:8443/auth/realms/ansible-
automation-platform
    attr_user_permanent_id: name_id
    attr_email: email
    attr_username: username
    attr_groups: groups
    url: https://sso.node:8443/auth/realms/ansible-
automation-platform/protocol/saml
    x509cert: "-----BEGIN CERTIFICATE-----CERTVALUE-----END
CERTIFICATE-----"
    attr_last_name: last_name
    attr_first_name: first_name
...
```

Once these settings are applied, log out of the Automation controller and click the person icon below the blue **Log In** button, circled in red in the following screenshot:

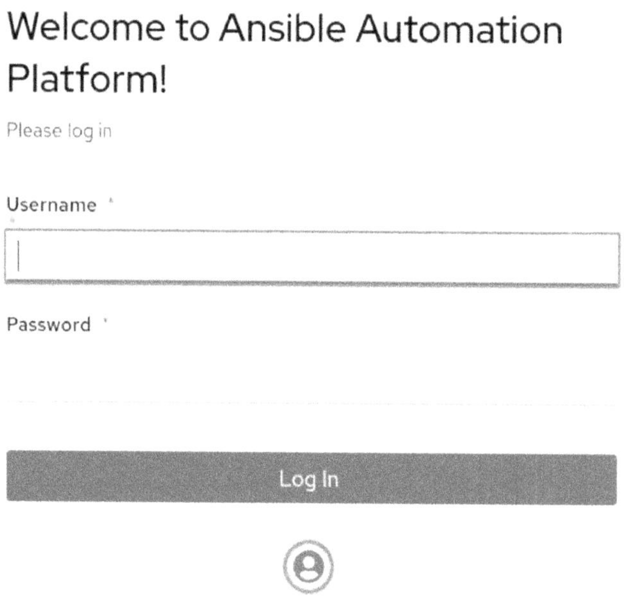

Welcome to Ansible Automation Platform!

Please log in

Username '

Password '

Log In

Figure 4.11 – Automation controller SSO button

Not every integration with the SSO server goes successfully, so the next section will cover a few options to diagnose issues that arise when integrating SAML.

Troubleshooting the SAML configuration

It can be difficult to diagnose issues with SAML authentication, but if you encounter any errors, the first place to look is `server.log`. This can be in a variety of places in a traditional SSO installation; however, for an Ansible Automation Platform installation, the default location is `/opt/rh/rh-sso7/root/usr/share/keycloak/standalone/log/server.log`.

Another useful tool is a SAML browser extension. For Firefox, there are two extensions to choose from. SAML Message Decoder will show the last SAML event, while SAML-tracer will show all events and show the **Extensible Markup Language** (**XML**) passed to the Automation controller.

SAML is not the only option to use with the SSO server—it is also possible to use LDAP to integrate with services such as Microsoft AD. The next section will cover using LDAP instead of SAML for enterprise authentication.

Integrating LDAP with Microsoft AD

LDAP integration is possible without the SSO server. The Automation controller will refer directly to the LDAP server. To configure it in the GUI, navigate to **Settings | LDAP authentication**. There are multiple tabs for LDAP as the controller can be configured to communicate with up to five different profiles.

A key point is the values to use; at the top of the previous section is a reference on how to get values to connect to an AD server. Those same values will be reused here.

As before, it is possible to configure these settings through the GUI, API, modules, and roles. There are instructions on how to use each earlier in this chapter.

Next, we look at the values used in an LDAP configuration. Each section also contains settings for the recommended basic setup for the same Microsoft AD server from the SSO section.

Automation controller LDAP bind and user variables

The following variables set bind and user settings for the controller:

GUI Field	API Variable	Description
LDAP Server URI	AUTH_LDAP_SERVER_URI	The server URI
LDAP Bind DN	AUTH_LDAP_BIND_DN	DN of LDAP Admin
LDAP Bind Password	AUTH_LDAP_BIND_PASSWORD	Password of LDAP Admin
LDAP User Search	AUTH_LDAP_USER_SEARCH	DN of user search
LDAP User DN Template	AUTH_LDAP_USER_DN_TEMPLATE	Alternative to DN user search
LDAP User Attribute Map	AUTH_LDAP_USER_ATTR_MAP	Map LDAP scheme to controller scheme

Figure 4.12 – Automation controller base LDAP variables

These are used in settings in the following format:

```
// ldap/aap_settings.yml
---
controller_settings:
 settings:
   AUTH_LDAP_SERVER_URI: LDAP://server2022.ansible.lcl
   AUTH_LDAP_BIND_DN:
CN=Administrator,CN=Users,DC=ansible,DC=lcl
   AUTH_LDAP_BIND_PASSWORD: "$encrypted$"
   AUTH_LDAP_USER_SEARCH:
   - CN=Users,DC=ansible,DC=lcl
   - SCOPE_SUBTREE
   - "(sAMAccountName=%(user)s)"
   AUTH_LDAP_USER_DN_TEMPLATE:
   AUTH_LDAP_USER_ATTR_MAP:
     first_name: givenName
     last_name: sn
     email: mail
...
```

Automation controller LDAP group variables

The following variables set group settings for the controller:

GUI Field	API Variable	Description
LDAP Group Search	AUTH_LDAP_GROUP_SEARCH	LDAP group search query to use
LDAP Group Type	AUTH_LDAP_GROUP_TYPE	Group type to use depending on server
LDAP Group Type Parameters	AUTH_LDAP_GROUP_TYPE_PARAMS	Group type parameters name and member attributes
LDAP Require Group	AUTH_LDAP_REQUIRE_GROUP	Set group membership requirements to limit to groups
LDAP Deny Group	AUTH_LDAP_DENY_GROUP	Deny login to groups set here

Figure 4.13 – Automation controller LDAP group variables

These are used in settings in the following format:

```
// ldap/aap_settings.yml
---
    AUTH_LDAP_GROUP_SEARCH:
    - CN=Users,DC=ansible,DC=lcl
    - SCOPE_SUBTREE
    - "(objectClass=group)"
    AUTH_LDAP_GROUP_TYPE: MemberDNGroupType
    AUTH_LDAP_GROUP_TYPE_PARAMS:
      member_attr: member
      name_attr: cn
    AUTH_LDAP_REQUIRE_GROUP:
    AUTH_LDAP_DENY_GROUP:
...
```

Automation controller LDAP miscellaneous variables

The following variables set the remaining LDAP settings for the controller:

GUI Field	API Variable	Description
LDAP User Flags By Group	AUTH_LDAP_USER_FLAGS_BY_GROUP	Set superusers and system admins by groups
LDAP Organization Map	AUTH_LDAP_ORGANIZATION_MAP	Maps to set organizations and admins
LDAP Team Map	AUTH_LDAP_TEAM_MAP	Maps to set teams inside an organization
LDAP Start TLS	AUTH_LDAP_START_TLS	Enable TLS when not using SSL

Figure 4.14 – Automation controller LDAP miscellaneous variables

These are used in settings in the following format:

```
// ldap/aap_settings.yml
---
    AUTH_LDAP_USER_FLAGS_BY_GROUP:
      is_superuser:
      - CN=ansible_admin,CN=Users,DC=ansible,DC=lcl
    AUTH_LDAP_ORGANIZATION_MAP:
      Default:
        users:
        - CN=ansible,CN=Users,DC=ansible,DC=lcl
        - CN=ansible_admin,CN=Users,DC=ansible,DC=lcl
        admins: CN=ansible_admin,CN=Users,DC=ansible,DC=lcl
        remove_admins: true
...
```

And these are the settings for the team map:

```
// ldap/aap_settings.yml
---
    AUTH_LDAP_TEAM_MAP:
      LDAP Engineering:
        organization: LDAP Organization
        remove: true
        users: cn=engineering,ou=groups,DC=ansible,DC=lcl
      LDAP IT:
        organization: LDAP Organization
        remove: true
        users: cn=it,ou=groups,DC=ansible,DC=lcl
      LDAP Sales:
        organization: LDAP Organization
        remove: true
        users: cn=sales,ou=groups,DC=ansible,DC=lcl
...
```

Additional information and references can be found here:

https://docs.ansible.com/ansible-tower/latest/html/administration/
ldap_auth.html

Troubleshooting

A useful tool to get information about authenticating LDAP is the `ldapsearch` tool. This uses the bind DN account and password with the −D option, and then will perform a search with the −b reference, as follows:

```
ldapsearch -x  -H ldap://server2022.ansible.lcl -D "
CN=Administrator,CN=Users,DC=ansible,DC=lcl " -w Password123 -b
"cn=Users,dc=ansible,dc=lcl"
```

This is used to get information back from the LDAP server to identify the information being returned for users, groups, and other objects.

In addition, there are two settings to set to get more information from the tower logs. The first is in `/etc/tower/conf.d/ldap.py`. If the file does not exist, create it, and add the following line of code:

```
LOGGING['handlers']['tower_warnings']['level'] =  'DEBUG'
```

After making this change, run the `automation-controller-service restart` command on the controller command line.

There is also a Automation controller setting located in the GUI. Navigate to **Settings | Logging settings | Edit | Logging Aggregator Level Threshold** and change this level to **DEBUG**. This can also be set as a setting via either the API, a module, or a role as `LOG_AGGREGATOR_LEVEL='DEBUG"`.

After these changes have been made, refer to the logs in `/var/log/tower/tower.log`.

Setting up other authentication methods

There are many other options for authentication available through either the Automation controller or the RH-SSO server. The general idea is the same, and it is best to consult a specific guide to do so. Unfortunately, because there are over a dozen methods, the most popular SAML and LDAP options were chosen for examples. More guidance can be found here: `https://docs.ansible.com/ansible-tower/latest/html/administration/ent_auth.html`.

There are instances where users and teams may need to be added manually to the Automation controller outside of an enterprise authentication system. The next section will go over how to add teams and users directly to the Automation controller.

Adding users and teams to the Automation controller without an IdP

The Automation controller is also able to have teams and users directly added to the system. With a development system or a small user base, this can be preferable to using an external authentication provider.

The following fields are used to define a user:

- username—Username for the user.
- password—Password for the user.
- email—Email address of the user.
- first_name—First name of the user.
- last_name—Last name of the user.
- is_superuser—Is an administrator.
- is_system_auditor—Is an administrator.
- update_secrets—True will always change the password if the user specifies the password, even if the API gives $encrypted$ for the password. False will only set the password if other values change too.
- groups—Groups user belongs to.
- append—Whether to append or replace the group list provided (modules and roles only).
- is_superuser—Is an administrator.

The following fields are used to define a team:

- name—Name for the team
- new_name—Role/module field to change the team name
- description—Description of the team
- organization—The organization the team belongs to

Adding users can be done in the GUI through the following steps:

1. Navigate to the controller via **Dashboard | Users | Add**.
2. Fill in the relevant fields—**Username**, **Password**, **User Type**, and **Organization** are required.
3. Click **Save**.

For teams, the process is similar:

1. Navigate to the controller via **Dashboard** | **Teams** | **Add**.

2. Fill in the relevant fields—**Name** and **Organization** are required.

3. Click **Save** | **Access** | **Add**.

4. Click **Save** | **Access** | **Add** | **Users**.

5. Select the users you wish to add to the team.

6. Click **Next**.

7. Choose which role to grant those users: **Admin**, **Member**, or **Read**.

8. Click **Save**.

Repeat as needed till all users and teams are added.

To use the module method, the following playbook would be used:

```
// controller/create_user_groups_module.yml
```

This is best for doing singular tasks to add users as a one-off task.

The user module invocation creates a superuser and sets their email, as follows:

```
- name: Create user
  ansible.controller.user:
    username: joe_123
    password: password123
    email: joe_123@example.org
    superuser: yes
```

Next, the team module creates a team in the default organization, like so:

```
- name: Create team
  ansible.controller.team:
    name: Joes
    description: All the Joes
    organization: Default
```

This will use the role module to add the user to the team as a member, as follows:

```
- name: Add jdoe to the member role of My Team
  ansible.controller.role:
    user: joe_123
    target_team: Joes
    role: member
...
```

Here's an example of using a configuration file to configure users and groups:

```
//controller/config/user_groups.yml
---
```

An account for the administrator joe_123 is created, as illustrated in the following code snippet:

```
controller_user_accounts:
  - user: joe_123
    is_superuser: false
    password: password123
```

A Joes team is created in the default organization, as follows:

```
controller_teams:
  - name: Joes
    organization: Default
```

The following code will add the user to the team as a member:

```
controller_roles:
  - user:  joe_123
    team: Joes
    role: member
```

Use the roles of users, teams, and roles to apply the configuration to the controller, as follows:

```
//controller/config/user_groups.yml
  pre_tasks:
    - name: Include vars from configs directory
      include_vars:
```

```
      dir: ./configs roles:
  - redhat_cop.controller_configuration.users
  - redhat_cop.controller_configuration.teams
  - redhat_cop.controller_configuration.roles
...
```

Users and teams that are added to the Automation controller do not cross over to the Automation hub. The process for adding users and teams to the Automation hub is different from that of the controller. The next section will cover how to add those users and teams.

Adding users and groups to the Automation hub without an IdP

The Automation hub can create its own users and groups on its own; however, it is built to use those set in the RH-SSO server. It is recommended to use the same methods discussed earlier to add IdPs to the SSO server. Users and groups cannot be created with the GUI; however, the module and role methods can be used to add them.

The following fields are used to define a user:

- username—Username for the user
- password—Password for the user
- groups—Groups user belongs to
- append—Whether to append or replace the group list provided (modules and roles only)
- first_name—First name of the user
- last_name—Last name of the user
- email—Email address of the user
- is_superuser—Is an administrator

The following fields are used to define a group:

- name—Username for the group
- permissions—Username for the group

User and groups can be added with modules. The following code snippet is an excerpt from the chapter's code repository:

```
// hub/create_user_groups_module.yml
---
```

The following task creates a group with the name Joes:

```
- name: Create group
  redhat_cop.ah_configuration.ah_group:
    name: Joes
    state: present
```

This task adds permissions to the Joes group using the ah_group_perm module, like so:

```
- name: Add group permissions
  redhat_cop.ah_configuration.ah_group_perm:
    name: Joes
    perms:
      - change_collectionremote
      - view_collectionremote
```

The ah_user module then creates an administrator for the Joes group, like so:

```
- name: Create User
  redhat_cop.ah_configuration. ah_user:
    username: joe123
    password: 123456789
    groups:
      - Joes


  . . .
```

The recommended way is to use roles.

CaC is still the recommended way to maintain configuration in the Automation hub by using a file to store information about a list of users and groups, as illustrated here:

```
//hub/configs/user_groups.yml
---
```

The following code creates an `ah_groups` group with a few permissions:

```
- name: Joes
  perms:
    - change_collectionremote
    - view_collectionremote
```

The following code creates an administrator that belongs to the `Joes` group:

```
ah_users:
  - username: joe123
    password: 123456789
    groups:
      - Joes
    is_superuser: true
...
```

This excerpt from the playbook imports the variable file from the `configs` directory and invokes `ah_configuration` roles that then apply the configuration:

```
//hub/create_user_groups_roles.yml
  pre_tasks:
    - name: Include vars from configs directory
      include_vars:
        dir: ./configs roles:
    - redhat_cop.ah_configuration.group
    - redhat_cop.ah_configuration.user
...
```

This should allow for the creation of accounts outside of authentication providers if needed, as well as promoting any users that need to be system administrators. More information about roles and permissions for users and teams can be found in *Chapter 6, Configuring Role-Based Access Control.*

Summary

This chapter covered how to integrate various methods of enterprise authentication and managing users and teams within the Automation controller and Automation hub. The enterprise authentication method sets up basic organizations but does not do anything more. Armed with how to configure settings, authentication providers, users, and groups on Ansible Automation Platform, it's time for the next step.

The next chapter will go over setting up the basics in the Automation controller and Automation hub. The basics include creating organizations and credentials, importing and exporting configuration in the controller, and adding content to the Automation hub.

Part 2:
Configuring AAP

Now that you have AAP installed, it is time to set up the various pieces of it, from organizations and inventories to job templates and workflows. You will learn how these pieces are created and managed and how they work together.

The following chapters are included in this section:

- *Chapter 5, Configuring the Basics after Installation*
- *Chapter 6, Configuring Role-Based Access Control*
- *Chapter 7, Creating Inventory, and Other Inventory Pieces*
- *Chapter 8, Creating Execution Environments*
- *Chapter 9, Automation Hub Management*
- *Chapter 10, Creating Job Templates and Workflows*

5

Configuring the Basics after Installation

With authentication set up, the next step is configuring the organizations and credentials inside the Automation controller. The organization represents how everything is organized. Everything from job templates to inventories is categorized by which organization they belong to. Within organizations are credentials, which are where secret variables are stored. Finally, once something is created in the controller, how do you get it out or, a better question, how is it best to manage organizations, credentials, and collections?

In this chapter, we're going to cover the following main topics:

- A dashboard introduction
- Creating organizations – the root of all objects in the controller
- Using credential types and credentials
- Importing and exporting objects in the controller

Technical requirements

This chapter will go over the platform and methods used in this book. All code referenced in this chapter is available at `https://github.com/PacktPublishing/Demystifying-the-Ansible-Automation-Platform/tree/main/ch05`. It is assumed that you have Ansible installed in order to run the code provided.

In addition, it is assumed that both an Automation controller and hub have been installed.

Dashboard introduction

The dashboard is the first thing that appears in the GUI when logging into the Automation controller. This will show the following things:

- The total number of hosts
- The number of failed hosts
- The number of inventories
- Inventory sync failures
- The number of projects
- Project sync failures
- A job-status graph of successful jobs versus failures
- Recent jobs
- Recent templates

The dashboard looks like this:

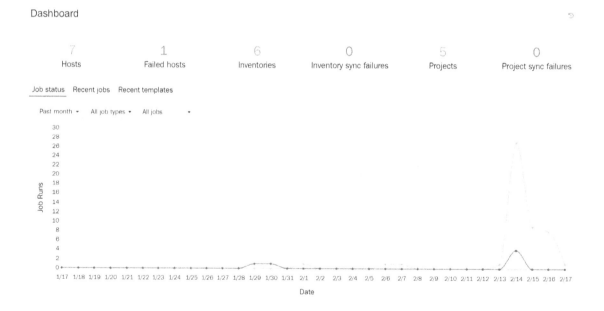

Figure 5.1 – The Automation controller dashboard

The dashboard chart can be adjusted to display a different period of time, only certain job types, or only successful or failed jobs.

Along the side is the navigation for accessing all the objects in the Automation controller:

Figure 5.2 – The Automation controller navigation bar

The navigation bar is the primary way of navigating the Automation controller.

Creating organizations – the root of all objects in the controller

Organizations are logical collections of objects in the Automation controller. Nearly all objects in the controller have some relation to an organization. It is the building block of everything in the Automation controller.

The following is a chart showing how organizations relate to the other objects in the Automation controller:

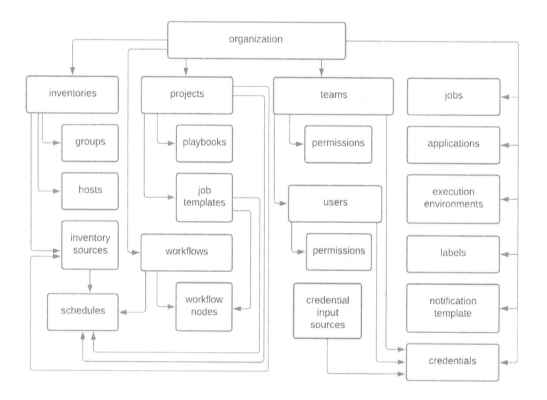

Figure 5.3 – An organizational relation chart

Organizations are also a delimiter in terms of access and uniqueness; if a user is a member of organization A, they are unable to use objects in organization B without becoming a member of organization B. There can be multiple job templates with the same name in a controller if they all belong to separate organizations.

> **Important Note**
> It may make more sense to use departments or different groups within a department as the basis for organizations so that each group can have its own credentials, job templates, and other objects separate from other teams. This allows each group to have more control over their content. How to divide groups and departments is entirely dependent on how the Automation controller will be used in the business.

Job templates are an exception to this rule. Job templates do not have an explicit organization. They inherit their organization from the project it is configured to use.

It is important when using API and module calls to reference the organization that an object belongs to. Without referencing the organization, it is very possible to have duplicate entries returned.

Creating an organization using the GUI

Organizations can be created in the GUI by using the following steps:

1. Navigate to **Organizations** on the right-hand sidebar.
2. Click on **Add** and fill in the following fields:

   - **Name**: The unique organization name.
   - **Description**: The description of the organization.
   - **Max hosts**: Defaults to **0** and can be set to divide up the number of hosts in a subscription.
   - **Instance Groups**: Which execution nodes the organization is allowed to run on.
   - **Execution environment**: The default execution environment to use when one has not been specified.
   - **Galaxy Credentials**: The galaxy credentials used to download collections and roles with; the order of galaxy credentials determines the order in which to use them. These are essential for where the controller pulls collections from for use.

3. Click **Save**.

Organizations can be created when setting authentication as discussed in *Chapter 4, Configuring Settings and Authentication*, although this method does not set any of the additional options. Use the previous method to update an existing organization created with authentication settings.

Creating an organization using modules

Organizations can be created using modules with the following playbook:

```
//organizations/organizations_module.yml
---
```

The same inputs for the name, credentials, and other elements discussed in the GUI can also be used in the module to make changes or updates inside a playbook:

```
- name: Change Organization
  ansible.controller.organization:
    name: Engineering
    galaxy_credentials:
      - Ansible Galaxy
...
```

This will create a single organization with the name Engineering and attach to it a galaxy credential for use in syncing collections in projects.

> **Important Note**
>
> When creating these, there is a chicken-and-egg dilemma. First, the organization must be created, then the credential can be created, and finally, the credential can be added to the organization. When creating or defining things for the first time, this is important; however, subsequent invocations to update or maintain the organization and credential do not have this problem.

Creating an organization using roles

Organizations can be created using roles with the following configuration file:

```
//organizations/organizations.yml
```

The roles have everything defined in configuration files. The `controller_organizations` variable takes a list of organizations and their variables to maintain organizations in the controller:

```
controller_organizations:
  - name: Satellite
    galaxy_credentials:
      - Ansible Galaxy
  - name: Default
```

```
    galaxy_credentials:
      - Ansible Galaxy
```

The `redhat_cop.controller_configuration.organizations` role takes the variable input to create organizations using the following role:

```
  - redhat_cop.controller_configuration.organizations
...
```

This will create multiple organizations and can be expanded to include as many as needed in the list, each with its own options.

Organizations are key for creating objects in the Automation controller. With this key building block in place, the next step is to create credential types and credentials.

Using credential types and credentials

Credentials are the way to store secrets and passwords inside of the Automation controller. What a credential stores and how it can be accessed in a playbook is dictated by what type it is. The controller has 26 built-in credential types, ranging from SSH keys to those designed to interact with secret management systems such as HashiCorp Vault. Some of the built-in credentials are also made to access services such as AWS or Azure.

Credential types

To understand credentials, it is important to look at the underlying credential types. Credential type definitions can be found by navigating to the web page (`https://{{controller_fqdn}}/api/v2/credential_types/`) of the controller. There is also a playbook to export the credential API definitions in this chapter's repository (`ch05/credentials/credential_types_names.yaml` and `ch05/credentials/credential_types.yaml`).

Custom credential types can be especially useful to store global sensitive variables used by playbooks/job templates. The long-accepted practice has been to use `ansible-vault` to store variables and a vault password as a credential in the controller to read them. However, it is just as practical to store these variables as custom credential types within the controller. Both practices are acceptable; however, it may be easier to maintain a credential stored in the Automation controller than a vaulted variable that can be in multiple repositories, projects, and playbooks, leading to difficulty in maintaining and updating the credential.

> **Important Note**
> Only administrators can create credential types. They are global for all of the Automation controllers, across organizations. Organization administrators cannot create credential types that are only accessible to their organization. Organization administrators can, however, create credentials based on a credential type that is only accessible to their organization.

Credential types are made up of two parts – inputs and injectors. The inputs determine what the user inputs to create a credential. The injectors determine what is accessible in the playbook.

The inputs are made up of a list of fields, such as the following:

```
// credentials/credential_type_fields.yml
---
fields:
 - id: password
   label: Password
   help_text: A password
   type: string
   choices:
     - A
     - B
     - C
   format: ssh_private_key
   multiline: false
   secret: true
required:
 - password
...
```

Each field is an entry for an input and has the following parts:

- `id`: The unique identifier for the field to be referenced by the injectors.
- `label`: The text description to be displayed in the GUI for the ID.
- `help_text`: Optional – helper text that is displayed by clicking **?** next to the label in the GUI to explain the field.
- `type`: A type of credential and either string or Boolean.
- `choice`: Optional – a list of values the field can be.
- `format`: Optional – a formatter for the string. It can only be `ssh_provkate_key` or `url`.
- `secret`: Optional – a Boolean. It says whether the field is a secret and should be encrypted, and its text is hidden in playbooks.

Once the fields are created, injectors use the fields to pass values onto the playbooks. These come in three categories – environmental variable, extra variable, and file. The extra variables and environmental variables are automatically injected into the playbook.

The injectors are made up of dictionaries such as the following:

```
// credentials/credential_type_injectors.yml
---
injectors:
  env:
    REST_USERNAME: "{{ rest_username }}"
    REST_PASSWORD: "{{ rest_password }}"
    MY_CERT_INI_FILE: "{{ tower.filename.cert_file }}"
    MY_KEY_INI_FILE: "{{ tower.filename.key_file }}"
  extra_vars:
    rest_username: "{{ rest_username }}"
    rest_password: "{{ rest_password }}"
  file:
    template.cert_file: |-
      [mycert]
      {{ cert }}
    template.key_file: |-
      [mykey]
      {{ key }}
...
```

Each category provides variables injected into the playbook. Both env and extra_vars use standard dictionary fields to then set a field ID as a Jinja variable substitution in double curly brackets, "{{ variable_name }}", as shown in the preceding file.

The file injector creates a file based on the template provided whose absolute path is stored in the tower.filename.* variable, where * is the key used for the template. This can then be used as an extra variable or environmental variable for use in the playbook. An example of this can be found in the preceding file.

> **Important Note**
>
> When creating credential types, it is important to avoid creating collisions in the injectors. No two injectors should use the same variable name. It is also important to avoid using the ANSIBLE_ or ansible_ prefix with either extra variable or environment variables to avoid collisions with the inner workings of the Ansible settings, as those prefixes are reserved by Ansible.

The two previous fields are used to create the custom credential type. This consists of the name, the kind of credential, and the input and injector fields previously described.

Each credential type has the following parts:

- **Name**: Required – the name of the credential type. It must be unique.

- **Descriptions**: Optional – a description of the credential type.

- **Kind**: Required – the kind of credential; the default field value is `cloud`. Setting this to any other value has been depreciated for custom credential types.

- **Inputs**: Required – input fields, as described previously.

- **Injectors**: Required – injector fields, as described previously.

The following file shows how this is used for each credential type:

```
// credentials/credential_type_combined.yml
...
name: REST API Credential
description: REST API Credential
kind: cloud
inputs:
injectors:
...
```

> **Important Note**
> The Automation controller has 26 built-in credential types. However, only three of those have injectors to be used in playbooks – Insights, Red Hat Virtualization, and Red Hat Ansible Automation Platform. The controller still uses these credentials, but they are not accessible by the playbook.

Credentials

Credentials are used for two purposes – connecting to services and injecting variables into playbooks. The vast majority of the built-in credential types are set to connect to services. The one thing in common is that they allow for authentication, hence the credential terminology.

The purpose of credentials is to store secrets. These secrets can be used to authenticate to services or store passwords. However, nothing prevents them from being used to store global variables for use in multiple job templates, even if there are no secrets being stored.

The next few subsections cover the different types of credentials.

Cloud inventory

These credentials are used for syncing cloud inventory with one of the following services:

- Amazon Web Services
- Google Compute Engine
- Insights
- Microsoft Azure Resource Manager
- OpenStack
- Red Hat Satellite 6
- Red Hat Virtualization
- VMware vCenter

These credentials set access to their service so that inventory hosts can be synchronized.

Automation controller credentials

The following are credentials that are related to the Ansible Automation Platform services:

- **Ansible Galaxy/Automation Hub API token**: Connect to Ansible Galaxy or your on-premises Automation hub.
- **Red Hat Ansible Automation Platform**:
  - Access credentials for an Automation controller
  - Uses injected variables to be picked up by the modules
  - **Vault**: A vault password and ID used to decrypt files or variables encrypted with `ansible-vault`.
- **OpenShift or Kubernetes API Bearer token**: Connect to instance groups that run job templates on containers.
- **Container registry**: Connect to a container registry to pull container images to run jobs on.

Source code repositories

These credentials are specifically for connecting to repositories to synchronize projects:

- **GitHub personal access token**: Connect to GitHub for syncing projects.
- **GitLab personal access token**: Connect to GitLab for syncing projects.
- **Source control**: Connect to a remote revision control system such as Git or Subversion.

Inventory credentials

These credentials are for connecting to inventory hosts through a variety of methods – username, password, SSH, and different escalation methods:

- Machine:

  - Connect to a machine using a username/password or SSH key.

  - Sets privilege escalation methods for this connection.

- Network:

  - Connect to networking devices using a username/password or SSH key.

  - Especially for network modules such as `netconf`.

  - More details can be found here: `https://docs.ansible.com/ansible/devel/network/user_guide/platform_index.html#platform-options`.

Secret lookup plugins

These are credentials for plugins to look up secrets in external systems. These differ in how they are used, but all facilitate looking up external secrets. More details on these lookup plugins can be found here: `https://docs.ansible.com/automation-controller/latest/html/userguide/credential_plugins.html#ug-credential-plugins`:

- CyberArk AIM Central Credential Provider Lookup

- Microsoft Azure Key Vault

- Centrify Vault Credential Provider Lookup

- CyberArk Conjur Secret Lookup

- HashiCorp Vault Secret Lookup

- HashiCorp Vault Signed SSH

- Thycotic DevOps Secrets Vault

- Thycotic Secret Server

A more complete breakdown of every input for these credentials can be found here: `https://docs.ansible.com/automation-controller/latest/html/userguide/credentials.html#ug-credentials`.

The creation of credential types and credentials using various methods

Credential types and credentials can be created through the GUI, API, modules, or roles. The next few sections will cover the different ways of creating them.

Creating credential types using the GUI

Credential types can be created in the GUI by taking the following steps:

1. Navigate to **Credential Types** on the right-hand sidebar.

2. Click on **Add** and fill in the following fields:

    *   **Name**: Required – the name of the credential type. It must be unique.

    *   **Descriptions**: Optional – a description of the credential type.

    *   **Inputs**: Required – input fields, as described previously.

    *   **Injectors**: Required – injector fields, as described previously.

3. Click **Save**.

Creating credentials using the GUI

Credential types with the `kind` value `cloud` can be created in the GUI by using the following steps:

1. Navigate to **Credentials** on the right-hand sidebar.

2. Click on **Add** and fill in the following fields:

    *   **Name**: Required – the name of the credential. It must be unique to your organization.

    *   **Descriptions**: Optional – a description of the credential.

    *   **Organization**: Required – the organization name.

    *   **Credential type**: Required – select the credential type from the list.

3. Fill in the fields as prompted for the credential type selected.

4. Click **Save**.

> **Note**
>
> For this and the following methods, Jinja `raw` formatting, `{% raw %}`, is used so that Ansible does not try and replace the variables using traditional methods.

Creating credentials and credential types using modules

Credentials and credential types can be created using modules by using the following playbook. This is simpler than using the API, as the organization and credential types do not need to be found beforehand and can be referenced by name:

```
// credentials/credentials_module.yml
---

    - name: Create Credential type
      ansible.controller.credential_type:
        name: REST API Credential
        description: REST API Credential
```

Notice that the `kind` value `cloud` is used. This is the default value used when a credential type is created in the GUI, and the module defaults to this value as well. However, there is no reason to use any other value in this field. It is a leftover artifact from a previous iteration of credential type creation.

```
        kind: cloud
        inputs:
          fields:
            - id: rest_username
              type: string
              label: REST Username
            - id: rest_password
              type: string
              label: REST Password
              secret: true
          required:
            - rest_username
            - rest_password
```

The injectors supply what is injected for the playbook to pick up, which includes environment variables or extra Ansible variables:

```
        injectors:
          env:
            rest_password_env: "{% raw %}{{ rest_password }}{%
    endraw %}"
            rest_username_env: "{% raw %}{{ rest_username }}{%
    endraw %}"
```

```
        extra_vars:
          rest_password: "{% raw %}{{ rest_password }}{%
endraw %}"
          rest_username: "{% raw %}{{ rest_username }}{%
endraw %}"
```

With the credential type created, a credential or multiple credentials can be created from it. The following example shows how the module works by providing the inputs:

```
  - name: Create Credentials
    ansible.controller.credential:
      name: REST API Credential
      description: REST API Credential
      credential_type: REST API Credential
      inputs:
        rest_username: rest_user
        rest_password: password
...
```

Creating credentials and credential types using roles

Credentials and credential types can be created using roles by using the following playbook:

```
// credentials/credentials_roles.yml
```

The controler_contredential_types variable is a list of credential types.   controller_credential_types:

```
      - name: REST API Credential
        description: REST API Credential
        kind: cloud
```

The input is set here just as in the module:

```
        inputs:
          fields:
            - id: rest_username
              type: string
              label: REST Username
            - id: rest_password
              type: string
```

```
            label: REST Password
            secret: true
      required:
        - rest_username
        - rest_password
```

The injectors are in a special format, like in the env module the raw Jinja delimitator is used; however, because of it being picked up by a role, a double space is needed between the first { { brackets. This is needed to make sure things are interpreted correctly by Ansible:

```
      injectors:
        env:
          rest_password_env: "{% raw %}{  { rest_password }}
{% endraw %}"
          rest_username_env: "{% raw %}{  { rest_username }}
{% endraw %}"
        extra_vars:
          rest_password: "{% raw %}{  { rest_password }}{%
endraw %}"
          rest_username: "{% raw %}{  { rest_username }}{%
endraw %}"
        file:
          template: '[ini_field]
            rest_username={  {rest_username}}
            rest_password={  {rest_password}}'
  controller_credentials:
    - credential_type: REST API Credential
      name: REST API Credential
      description: REST API Credential
      inputs:
        rest_username: rest_user
        rest_password: password
      organization: Satellite
```

Using the previous variables with `credential_types` and credentials roles will add the configuration to the controller:

```
roles:
    - redhat_cop.controller_configuration.credential_types
    - redhat_cop.controller_configuration.credentials
...
```

> **Important Note**
>
> For this, Jinja `raw` formatting, {% `raw` %}, is used so that Ansible does not try and replace the variables using traditional methods. In addition, a double space ({   }) is used between brackets. This is a workaround, as when importing a file, Ansible will try and force the replacement of the variable. It is important to use double spacing when using the roles for injectors.

With the organization, credential types, and credentials created, the next topic explores importing and exporting objects into the Automation controller.

Importing and exporting objects into the controller

With objects created inside of the controller, it is important to be able to export and import those objects. This can be used to back up and restore some aspects of the Automation controller as well. However, this method has limitations on what it can export.

I'd personally recommend using **Configuration as Code (CaC)** to manage objects on Ansible Automation Platform. Users with existing deployments need a way to move objects to definition files in order to move to a CaC model. Using the export methods described in this section facilitates the move to CaC.

The modules used to export require an underlying Python module, which can be installed using the following command:

```
$ pip install awxkit
```

This allows for either the `awx.awx` or `ansible.controller` collection `import` and `export` modules to act on objects from the API.

Currently, this is limited to the following controller objects:

- `applications`
- `credential_types`
- `credentials`
- `execution_environments`

- inventory
- inventory_sources
- job_templates
- notification_templates
- organizations
- projects
- teams
- users
- workflow_job_templates

The caveats of these objects are as follows:

- Workflow job template approval nodes are not exported. If using approval nodes, these will need to be added manually.

- Credentials will not export secret values. There is currently not a good solution for exported protected secrets.

Now that we understand the limitations of the export, it is time to export the objects to files that can be used for import:

```
// export_import/export.yml

The export module can export specific objects or all objects.
- name: Export workflow job template
      ansible.controller.export:
        all: True
      register: export_results

   - debug:
       var: export_results

   - name: Export to file
     copy:
        content: "{{ export_results | to_nice_yaml(indent=2)
}}"
        dest: configs/controller.yaml
...
```

This file can then be used to invoke the import of assets with the following tasks:

```
// export_import/import_module.yml
---
- name: Include vars from configs directory
  include_vars:
    file: controller.yaml

- name: Import all assets from our export
  ansible.controller.import:
    assets: "{{ assets }}"
...
```

The `redhat_cop.controller` configuration roles can also be used to import the objects:

```
$ ansible-playbook redhat_cop.controller_configuration.
configure_controller.yml --extra-vars "controller_configs_
dir=Demystifying-Ansible-Automation-Platform/ch05/export_
import/configs"
```

Here, `controller_configs_dir` is the folder that contains the configuration files.

This has covered the Automation controller importing and exporting tools to move to CaC.

Summary

This chapter covered the dashboard, the hub of navigation for the controller. In addition, we covered creating organizations to which most objects in the controller are related. Next, we covered credential types and credentials, which allow for keeping secrets or connecting to various services.

The next topic covered the import and export options that allow for better management of the objects inside of the controller. All these pieces form the building blocks upon which jobs work inside of the Automation controller.

The next chapter covers role-based access control of the different pieces after they are created.

6

Configuring Role-Based Access Control

Role-Based Access Control (RBAC) permissions are used to delegate access in both the Automation controller and hub. By default, system administrators have access to create and edit everything; however, in most environments, access should be limited to the minimum permissions needed to run. To achieve this, there are a variety of roles designed to grant that permission.

This chapter is broken up into the following sections:

- Assigning RBAC to the Automation controller
- Assigning RBAC to the Automation hub

Each of these sections covers its respective service. In each section, the different roles that can be granted will be covered, along with how to grant those roles to users; finally, each section will go into brief detail about how to use the roles practically for each service.

Technical requirements

This chapter will go over the platform and methods used in this book. All code referenced in this chapter is available at `https://github.com/PacktPublishing/Demystifying-the-Ansible-Automation-Platform/tree/main/ch06`. It is assumed that you have Ansible installed in order to run the code provided.

In addition, it is assumed that both an Automation controller and hub have been installed. This chapter will also go into setting RBAC with the hub and controller.

Assigning RBAC to the Automation controller

During the previous chapters, organizations, teams, and users were created. However, unless the users are admins, they have limited access inside the Automation controller.

Roles in the Automation controller follow the basic rules of RBAC . This means that each role has a set of permissions for what it is allowed to do. These roles fall into categories of allowing objects to be viewed, edited, and executed with the controller. In addition, the roles themselves are attached to organizations, teams, and users. Teams and users are interchangeable; it's just that teams are a collection of users.

In addition, there is a hierarchy of roles. The hierarchy of roles follows inheritance. Some roles inherit permissions from other roles. The way to think of this is that a read role can only read, a user role can read and use an object, and an admin role can edit, use, and read an object.

The roles available in the Automation controller and their corresponding module options are the following:

Module option	System role	Has permision to
admin	Admin role for a given object	Superuser privileges for a given Organization, Team, Inventory, Project, or Job Template
read	Read role for a given object	For an Organization, Team, Inventory, Project, or Job Template, be able to read all aspects of that object
member	Member role for a given object	Gives user access as a member of a given Organization or Team
execute	Execute role – Job Templates	Launch Job Template to create a Job
adhoc	Ad Hoc role – Inventory	Run commands on Inventory hosts
update	Update role – Project, Inventory	Update the Project or Inventory
use	Use role – Credential, Inventory, Project	Assign the given Credential to an Inventory, Project, or Job Template
approval	Approval role – Workflow Level	Approve an approval node in a workflow
auditor	Auditor role – All	Read Everything for a given Organization, Project, Inventory, or Job Template

Figure 6.1 – Object roles

These roles can be applied to a user or team to grant authority for doing individual actions. An admin has all the privileges below it.

Job templates are made up of specific parts, a project that contains the playbook, the inventory to run on, and it may or may not have a credential. Even if a user is assigned admin privileges on a job template, they may not have access to the underlying project, inventory, or credential. In order to add one of these to a given job template in editing, they must have use privileges for the item to be added.

To have *full* access to change everything, the user or team would need to have full admin access to the job template, project, inventory, and credential. The same pertains to projects that need credentials as well when editing them.

`use`, `update`, and `execute`, and `adhoc` are different, and fall under a **usage** category. The user does not need to have access to the underlying objects to have usage of them.

The auditor role is special, as it is an elevated read role, but it is specifically provided for denoting a user or team that would need full read access but not any other access. It also works well for a service account that is specifically for reading logs and other Tower information that should not have other access.

In addition to the previous roles, there are organizational admin roles:

Module option	System role	Manages all aspects of
project_admin	Admin for a project	Projects
inventory_admin	Admin for a inventory	Inventories
credential_admin	Admin for a credential	Credentials
workflow_admin	Admin for workflow	Workflows
notification_admin	Admin for a notification	Notifications
job_template_admin	Admin for a job template	Job templates
execution_environment_admin	Admin for an execution enviroment	Execution environments
N/A	System administrator	System
N/A	System auditor	Views all aspects of the system

Figure 6.2 – Admin roles

These roles give admin control over all objects of the given type for an organization. Just short of system administrators, it gives the user or team full access to all objects of that type in an organization.

> **Important Note**
> Systems roles such as auditor and administrator can only be applied to users using the user module and API endpoint, not with the role methods discussed in this section. Refer to *Chapter 4, Configuring Settings and Authentication*, on how to set those.

Applying these roles consists of three parts:

- Who is being given permission?
- To what object is the permission being granted?
- What role is being applied?

Navigate to the organization, team, or user in the GUI that you wish to edit access for.

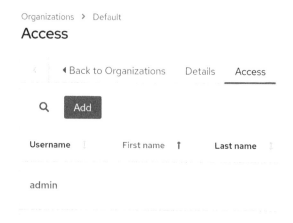

Figure 6.3 – Adding access to an organization

For organizations, navigating to **Organizations** | organization name | **Access** will open an interactive tool to apply various roles to users and teams. The same can be done for users or teams by navigating to the corresponding user or team and selecting **Roles** | **Add**.

To give admin permission to an inventory for a user, using a module, the `roles/set_role_using_module.yml` playbook would be used. The following gives the user administrator rights over the given inventory:

```
- name:  Set user to admin over inventor
  ansible.controller.role:
    user: joe123
    role: admin
    inventories:
      - Demo Inventory
...
```

Using the module can be especially useful for creating dynamic or new objects in the Automation controller and granting granular access to them immediately. However, using **Configuration as Code** (**CaC**) to maintain access control is the recommended method.

Finally, using the `redhat_cop.controller_configuration` roles that implement the module, it's easier to loop over the module and define the definitions as code. The following two list items give the `satlab` team and the `joe123` user admin permissions to a project and a job template respectively:

```
// roles/set_role_with_roles.yml
---
```

```
controller_roles:
  - projects:
      - Demo Project
    job_templates:
      - Demo Job Template
    team: satlab
    role: admin
  - projects:
      - Demo Project
    job_templates:
      - Demo Job Template
    user: joe123
    role: admin
```

The `roles` role in the controller configuration applies the definitions from the preceding code to the Automation controller:

```
roles:
  - redhat_cop.controller_configuration.roles
...
```

The role and module-based methods work well for maintaining the granular access for RBACs. One permission that is commonly overlooked is the approval permission for workflows.

Approval permissions for workflows

The details on approval nodes will be covered in the chapter about workflows; however, the approval role has its own place in RBAC. While those who can create and edit workflows have the right to create approval nodes and those with read access to workflows can see approvals, those with execute permissions on a workflow do not automatically have the ability to approve or deny approvals; that permission must be granted to the user or group. More details about workflow approvals can be found in *Chapter 10, Creating Job Templates and Workflows*, and here: `https://docs.Ansible.com/Ansible-tower/latest/html/userguide/workflow_templates.html?#approval-nodes`.

For those workflows where approvals are merely for a user to review before continuing, or where the approval is used to pause a workflow, workflow approval can be granted. In other case, the executing user may not be the one doing the approving but another team instead.

It is important to remember when using approval nodes that extra permissions must be granted to users to take advantage of them.

With the RBAC options covered, how can you practically set a system to govern access?

Practically using the RBACs

With the Automation controller, users fall into two camps – those who need to edit and those who need to run. If running CaC and being strict about it is the only way to change configuration, admin privileges can be locked to service accounts to make changes. For Enterprise users running jobs, the three things really needed to run are executing privilege on job templates that are not in workflow, workflows, and approval rights for the correct teams, and nothing more.

For developers doing testing and the tweaking of job templates and workflows, it is often easier to have them in separate organizations with admin control over everything, or even a fully separate development Automation controller instance that they have full systems admin access to.

In the end, the controls can be set up to be permissive or strict, whichever the administrators prefer.

Assigning RBAC to the Automation hub

Automation hub RBACs work much like the Automation controller. There are a few categories of roles, each granting access to a different aspect of the Automation hub. The categories are as follows:

- User management.

- Group management.

- Collection namespace: Collections are named with two parts, such as `redhat_cop.ah_configuration`. The first half before the dot is the namespace – in this case, `redhat_cop`.

- Collection management: Used to manage collections that are not part of remote collections.

- Remote collection management: Used to manage remote collections from either of the following:

  - Red Hat certified collections

  - Ansible Galaxy

  - Container registry management: Used to manage the container registry.

  - Task management.

Groups with these permissions can manage user configuration and access:

Permission	Role/Module Variable Name
Add user	add_user
Change user	change_user
Delete user	delete_user
View user	view_user

Figure 6.4 – User roles

Groups with these permissions can manage group configuration and access:

Permission	Role/Module Variable Name
Add group	add_group
Change group	change_group
Delete group	delete_group
View group	view_group

Figure 6.5 – Group roles

The following are the namespace permissions; groups with these permissions control collection namespaces:

Permission	Role/Module Variable Name
Add namespace	add_namespace
Change namespace	change_namespace
Upload to namespace	upload_to_namespace
Delete namespace	delete_namespace

Figure 6.6 – Namespace roles

Groups with the `modify_ansible_repo_content` permission can move content between repositories using the approval feature. They can certify or reject features to move content from the staging, which is where collections sit pending approval, to published or rejected repositories:

Permission	Role/Module Variable Name
Modify Ansible repository content	modify_ansible_repo_content
Delete collection	delete_collection

Figure 6.7 – Collection roles

Groups with these permissions can configure remote repository syncing from community or Red Hat certified sources under repository management:

Permission	Role/Module Variable Name
Change remote collection	change_collectionremote
View remote collection	view_collectionremote

Figure 6.8 – Remote collection roles

Groups with these permissions can manage the container registry:

Permission	Role/Module Variable Name
Change container namespace permissions	change_containernamespace_perms
Change containers	change_container
Change image tags	change_image_tag
Create new containers	create_container
Push to existing containers	push_container
Delete container repository	delete_containerrepository

Figure 6.9 – Container roles

Groups with these permissions can manage tasks:

Permission	Role/Module Variable Name
Change task	change_task
Delete task	delete_task
View all tasks	view_task

Figure 6.10 – Task management roles

> **Important Note**
>
> The **Super user** permission can only be applied to users using the GUI, the user module, and API endpoint, not with the role methods discussed in this section. Refer to *Chapter 4, Configuring Settings and Authentication* for how to set that permission.
>
> However, with the module and role, an additional permission exists of either '*' or 'all' to assign all the preceding permissions to a user without granting them system administrator status.

Setting roles in the GUI

Roles in the GUI are done for each group on a single page. The process is relatively easy.

To set roles in the GUI, use the following steps:

1. Navigate to and click on **User Access** on the left-hand side, and then click on **Groups**.

2. Click on the name of a group. Click **Edit**.

3. For each drop-down menu, select the role to grant.

4. Click **Save**.

This is what the GUI should look like:

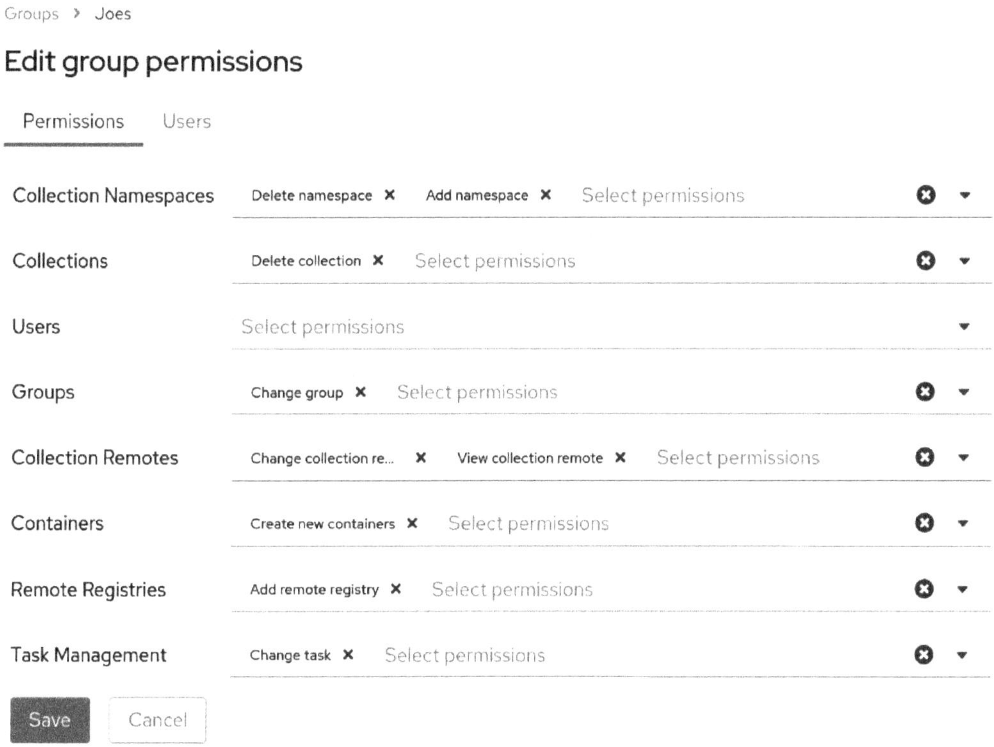

Groups > Joes

Edit group permissions

Permissions   Users

Collection Namespaces	Delete namespace ✕ Add namespace ✕ Select permissions	⊗ ▾
Collections	Delete collection ✕ Select permissions	⊗ ▾
Users	Select permissions	▾
Groups	Change group ✕ Select permissions	⊗ ▾
Collection Remotes	Change collection re... ✕ View collection remote ✕ Select permissions	⊗ ▾
Containers	Create new containers ✕ Select permissions	⊗ ▾
Remote Registries	Add remote registry ✕ Select permissions	⊗ ▾
Task Management	Change task ✕ Select permissions	⊗ ▾

Save   Cancel

Figure 6.11 – Automation hub roles

The steps for creating and giving permissions for a group with the modules are as follows. The group_perm module sets two permissions for the Joes group:

```
// hub/create_groups_module.yml
---
  - name: Add group permissions
    redhat_cop.ah_configuration.ah_group_perm:
      name: Joes
      perms:
        - change_collectionremote
        - view_collectionremote
      state: present
      ah_host: https://ah.node
      ah_username: admin
      ah_password: secret123
```

```
        ah_path_prefix: galaxy
        validate_certs: false
  . . .
```

The steps for creating and giving permissions for a group with the roles are as follows. The group role takes a list of groups with set permissions and applies them to the Automation hub:

```
// hub/create_groups_roles.yml
---
    ah_groups:
      - name: Joes
        perms:
          - change_collectionremote
          - view_collectionremote
```

This is the full name of the group role to apply permissions to a group:

```
    - redhat_cop.ah_configuration.group
  . . .
```

Knowing how to assign roles, how do you go about practically assigning access?

Practically using the RBACs

With the Automation controller, users are more likely to need to access various objects in order to be able to run jobs. In the Automation hub, most users likely just need view/read access in order to see remote collections and by default can see namespaces and containers. This allows them to see what is available to use in the controller. By extension, the Automation controller itself only needs view access to the hub as well.

Only those users with the need to upload and manage namespaces and push and create execution environments need to really have extra permissions on the Automation hub. This can vary depending on how strict or permissive your organization wishes to be.

Summary

This chapter covered what roles are available to use on the Automation controller and hub. It also covered how to go about assigning those roles to users' teams and groups.

Now that access for those using the services is covered, the next chapter will focus on the inventories of hosts, how to maintain them, and the creation of inventory plugins. These define what the Automation controller acts on.

7

Creating Inventory, and Other Inventory Pieces

Inventories describe what is being acted upon in the Automation controller. Job templates target hosts and groups that reside in inventories. Almost every company has a different way of storing the **source of truth** (**SOT**) when it comes to an inventory of devices. This chapter will go into the details of managing inventories, from adding them from files to interacting with outside services to populate the host and groups that the controller references.

In this chapter, we're going to cover the following topics related to inventories:

- Creating an inventory
- Using inventory sources
- Using base Ansible inventory plugins
- Using built-in automation controller inventory plugins
- Using other popular inventory plugins
- Writing your own inventory plugin

Technical requirements

This chapter will go over the platform and methods used in this book. All code referenced in this chapter is available at `https://github.com/PacktPublishing/Demystifying-the-Ansible-Automation-Platform/tree/main/ch07`. It is assumed that you have Ansible installed in order to run the code provided.

Creating an inventory

Inventories are where hosts are stored in the Automation controller. A host is anything that a task in a job template or playbook can act upon. It can be a Linux machine, a Windows **virtual machine (VM)**, a router, or even something that is accessible only with a **HyperText Transfer Protocol (HTTP) application programming interface (API)** endpoint. A collection of hosts is known as a group. A host doesn't need to belong to a group, but it can belong to multiple groups. In addition, inventories, groups, and hosts can also have variables attached that can be picked when a job is run. Each inventory is separate from other inventories, and only one can be used in a job at a time.

A feature of inventories is that they can be populated with sources. These sources range from **INItialization (INI)** or **YAML Ain't Markup Language (YAML)** files to Python plugins. They can populate groups, hosts, and variables as well. This is the recommended way of populating inventories. You can see an inventory overview in the following screenshot:

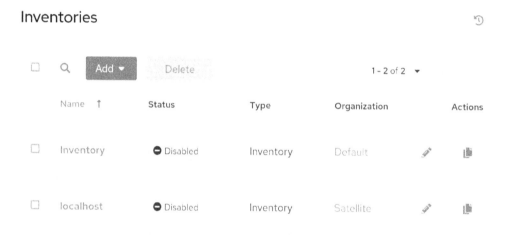

Figure 7.1 – Inventory overview on automation controller

The inventory page can be broken down into the following areas:

- **Name**: Name of the inventory.
- **Status**: Status of the last inventory synchronization. Disabled for inventories without sources.
- **Type**: Identifies whether an inventory is a standard inventory or a smart inventory.
- **Organization**: Name of the organization that the inventory belongs to.
- **Actions**: The following actions can be taken from this screen:
  - **Edit**: Edit an existing inventory.
  - **Copy**: Copy an existing inventory.

- **Add**: Add a new inventory.

- **Delete**: Delete selected inventories.

To start understanding inventories, let's begin by going over ways to create and populate an inventory without a source.

Creating and populating an inventory using the GUI

Inventories can be created in the **graphical user interface** (**GUI**) by using the following steps:

1. Navigate to **Inventories** on the right-hand sidebar.
2. Click on **Add | Inventory** and fill in the following fields:

   - **Name**: Unique inventory name inside the organization to which it belongs.

   - **Description**: A description of the inventory.

   - **Organization**: The organization to which the inventory belongs.

   - **Instance Groups**: Which execution nodes the inventory is allowed to run on. The order set is the order of precedence.

   - **Variables**: A set of variables can be entered here that will be applied to all hosts in the inventory.

3. Click **Save**.

It should look like this when complete:

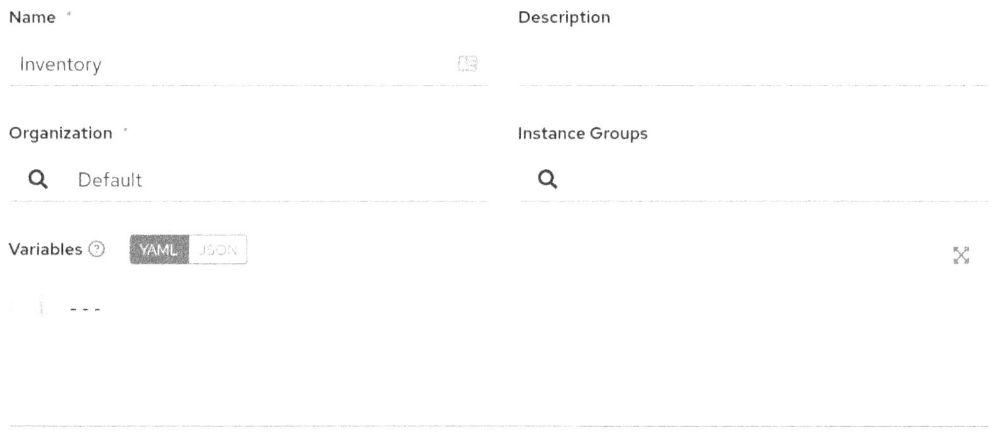

Figure 7.2 – Automation controller inventory creation

Inventories on their own do not work well. Hosts are needed to make an inventory useful. Hosts describe the machine and direct Ansible on how to connect to it.

Hosts can be created in the GUI by using the following steps:

1. Navigate to **Inventories** on the right-hand sidebar.

2. Click on the inventory to which you wish to add a host.

3. Click on the **Hosts** tab on the inventory description page.

4. Click on **Add** and fill in the following fields:

    • **Name**: Unique hostname inside the inventory to which it belongs.

    • **Description**: A description of the host.

    • **Variables**: A set of variables can be entered here that will be applied to the hosts in the inventory.

5. Click **Save**.

When complete, it should look like this:

Figure 7.3 – Automation controller host creation

Using hosts on their own is unwieldy, so it is possible to add them to groups. Groups are a logical collection of hosts and other groups.

Groups can be created in the GUI by using the following steps:

1. Navigate to **Inventories** on the right-hand sidebar.

2. Click on the inventory to which you wish to add a group.

3. Click on the **Groups** tab on the inventory description page.

4. Click on **Add** and fill in the following fields:

- **Name**: Unique group name inside the inventory to which it belongs.

- **Description**: A description of the group.

- **Variables**: A set of variables can be entered here that will be applied to the hosts in the group.

5. Click **Save**.

The creation of a group looks exactly like the creation of a host.

A group can contain other groups. This can be achieved by using the following steps:

1. Navigate to **Inventories** on the right-hand sidebar.

2. Click on the inventory to which you wish to add a group.

3. Click on the **Groups** tab on the inventory description page.

4. Click on the group name that you wish to add a group to.

5. Click the **Related Groups** tab on the **Group** page.

6. Click **Add** and then proceed as follows:

- Click **Add Existing Groups**: Select groups to add.

- Click **Add a New Group**: To add a new group and follow the same options for creating a group from before.

7. Click **Save**.

Though adding a host with the GUI might be simple for a one-off addition, it may not be practical. Another method for manipulating inventories is with modules.

Creating and populating an inventory with modules

A good way to add a host to an inventory during a playbook or Automation controller job is with modules. This can be useful when using workflows where a host is created and configured in one job, and then acted upon again later.

Inventories, hosts, and groups can be created using modules with the following excerpt from the `inventory_creation/aap_inventory_module.yml` playbook. The first step is to make sure the inventory exists in the correct organization:

```
- name: Add inventory
  ansible.controller.inventory:
    name: Basic Inventory
```

```
description: Create Inventory once piece at a time
organization: Default
```

The next step is to add a host and any applicable variables. This can also be useful for adding variables to a host for use in a later job. The code is illustrated in the following snippet:

```
- name: Add host
  ansible.controller.host:
    name: localhost
    description: "Local Host Group"
    inventory: Basic Inventory
    variables:
      hosts_var: some_val
```

Adding a host to a group is also possible by executing the following code:

```
- name: Add group
  ansible.controller.group:
    name: Cities
    inventory: Basic Inventory
    hosts:
      - basic_host
    preserve_existing_hosts: True
...
```

> **Important Note**
>
> The module and role have options for `preserve_existing_hosts` and `preserve_existing_children`. These module options are important as the request made to the automation controller API does not preserve hosts or child groups. Using this option will merge any new hosts or child groups with existing ones.

This is a great method for creating a host in a cloud, and then making sure that it is added to an inventory straight away, or for the next job in a workflow. However, it is not practical when trying to maintain a collection of hosts; a role is best suited for this.

Creating and populating an inventory with roles

When configuring Automation controller without an external inventory source, such as for a lab, one of the best ways to maintain an inventory is through **configuration as code** (**CaC**). The redhat_cop. controller configuration offers a way to do this.

Inventories, hosts, and groups can be created using roles with the inventory_creation/ configs/inventory.yml configuration file. The controller_inventories variable will take a list of inventories to populate on the automation controller, as illustrated in the following code snippet:

```
controller_inventories:
  - name: Basic Inventory
    description: Create Inventory once piece at a time
    organization: Default
    variables:
      inventory_var: inventory_value
```

A list of hosts can be added as well, like this:

```
controller_hosts:
  - name: basic_host
    inventory: Basic Inventory
    variables:
      hosts_var: some_val
```

Groups can also be defined in the configuration file. As noted previously, ensure the use of preserve variables to ensure that existing hosts in the group are not removed. The code is illustrated in the following snippet:

```
controller_groups:
  - name: Basic Group2
    inventory: Basic Inventory
  - name: Basic Group
    inventory: Basic Inventory
    variables:
      groups_variable: group_val
    hosts:
      - basic_host
```

```
    children:
      - Basic Group2
  ...
```

The following roles take the variable input to create inventories, hosts, and groups. This is an excerpt from the following file:

```
//inventory_creation/aap_inventory_role.yml
 roles:
   - redhat_cop.controller_configuration.inventories
   - redhat_cop.controller_configuration.hosts
   - redhat_cop.controller_configuration.groups
```

The three methods discussed—the GUI, module, and role—are all viable methods of updating the inventory depending on the circumstances. However, it does not replace having a **single SOT (SSOT)** for the inventory of hosts, groups, and their variables. Inventory sources are ways to manage Ansible and Automation controller inventories using external SSOT services. These services rely on inventory plugins to function.

Using inventory sources

This section will go into how to use sources inside of the Automation controller. These populate groups, hosts, and group/host variables in the inventory.

An inventory can have multiple inventory sources that can each contribute to the hosts and groups for that inventory.

> **Important Note**
>
> If the inventory plugin is dependent on a collection, that collection must be installed in the **execution environment** (**EE**) that the inventory source runs on. EEs are covered in the next chapter. It is possible to work around this by installing the collection to the project that the inventory source uses.
>
> Projects have also not been discussed and will be covered in *Chapter 10, Creating Job Templates and Workflows*.

The next three sections will go into detail about how to add inventory sources to an inventory in the Automation controller. Each section will cover a different method. The three methods covered will be using the GUI, modules, and Ansible roles.

A prerequisite to using an inventory source is a Git repository with a file for Ansible to read. In the following *Using base Ansible inventory plugins section*, we will cover this in more detail, but for now, an example inventory can be found in ch07/basic_inventory, as illustrated in the following code snippet:

```
# fmt: ini
[local_servers]
localhost ansible_connection=local

[local_servers:vars]
things=stuff

[all:vars]
admin_password=''
```

This is a basic Ansible inventory for example purposes. Keep it in mind for the next sub section, which covers how to create inventory sources.

Creating and synchronizing an inventory source with the GUI

The GUI can be used to quickly add an inventory source or make changes to an inventory source. This section will cover the details of the process and what each field is used for. An inventory can be created in the GUI by using the following steps:

1. Navigate to **Inventories** on the right-hand sidebar.

2. Click on the inventory to which you wish to add a group.

3. Click on the **Sources** tab on the inventory description page.

4. Click on **Add** and fill in the following fields:

   • **Name**: Unique group name inside the inventory to which it belongs.

   • **Description**: A description of the inventory source.

   • **Execution Environment**: The EE for the inventory source update to run on.

   • **Source**: The source of the inventory source. Choose a built-in inventory source.

   • Choose a **Credential** value if needed. Several options will appear depending on the source; they have been covered previously in this chapter.

- Choose a project that contains the definition file and fill in the following fields:

    - Choose a **Source Project** type to source the inventory definition file.

    - Type in the **Source Path** value of the inventory file, such as `inventories/plugin.yml`.

    - Choose a **Credential** value if needed.

- **Source Variables**: A set of variables can be entered here that will be used when running the inventory plugin.

- **Enabled Variable**: A variable to use to determine if the host is enabled.

- **Enabled Value**: Value when the host is considered enabled.

- **Host Filter**: If used, only hosts that match this **regular expression** (**regex**) will be imported.

- **Overwrite**: Delete child groups and hosts if they are not found in the latest source update.

- **Overwrite variables**: Override variables on every source update.

- **Cache Timeout**: If an inventory is set to update, an update is considered fresh for x seconds.

- **Update on launch**: Refresh the inventory when a job is run. Can impede other jobs.

- **Update on project update**: Boolean. If using a project source, if the project is updated, update the source.

- **Verbosity**: Verbosity level to run update **0-2**.

5. Click **Save**.
6. Click **Sync** to have the inventory source run.

It should look like this when complete:

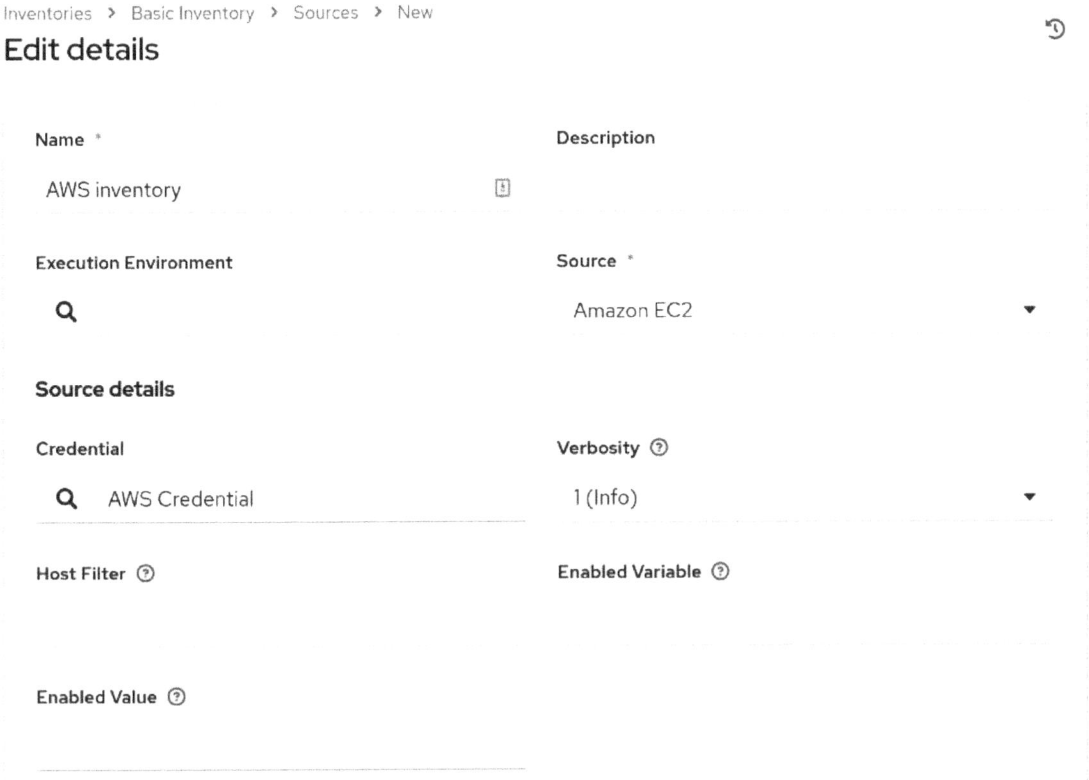

Inventories > Basic Inventory > Sources > New

Edit details

Name *

AWS inventory

Description

Execution Environment

🔍

Source *

Amazon EC2

Source details

Credential

🔍 AWS Credential

Verbosity ⑦

1 (Info)

Host Filter ⑦

Enabled Variable ⑦

Enabled Value ⑦

Figure 7.4 – Automation controller Amazon Web Services (AWS) inventory source creation

Using the GUI is the simplest and quickest way to add or change an inventory source; however, modules allow for them to be updated within a playbook or job.

Creating and populating an inventory source with modules

A good way to add or update an inventory source during a playbook or Automation controller job is with modules.

Inventory sources can be created using modules with the following excerpt from the `inventory_sources/aap_inventory_module.yml` playbook.

The example file makes sure that the inventory and project exist before creating an inventory source. This is no different than the previous inventory creation.

The important pieces of the inventory source to use are the source project, `source_path` variable, and inventory. The `scm` source type is available from the list of types at the beginning of this section and is the type to use for all projects. All options discussed in the GUI section can also be applied here. The code is illustrated in the following snippet:

```
- name: Add inventory source
  ansible.controller.inventory_source:
    name: Public APIs
    source: scm
    source_project: Inventory example
    source_path: inventories/plugin.yml
    inventory: Basic Inventory
    overwrite: true
```

To synchronize the project source, run the `inventory_source_update` module, providing the name of the source and inventory. The following code snippet illustrates this:

```
- name: Update a single inventory source
  ansible.controller.inventory_source_update:
    name: Public APIs
    inventory: Basic Inventory
    organization: Default
...
```

This can be used to create an initial inventory source or to update an inventory source. However, the best way to maintain the configuration is with roles.

Creating and populating an inventory source with roles

One of the best ways to maintain configuration for the Automation controller is with CaC. The `redhat_cop.controller` configuration offers a way to do this.

Inventory sources can be created using roles. The next code snippet is an excerpt of the `inventory_sources/config/inventory.yml` configuration file.

The configuration file contains configuration for the project and inventory as well.

The `controller_inventory_sources` variable will take a list of inventories to populate on the automation controller.

The code is illustrated in the following snippet:

```
controller_inventory_sources:
  - name: Public APIs
    source: scm
    source_project: Inventory example
    source_path: inventories/plugin.yml
    inventory: Basic Inventory
    organization: Default
    overwrite: true
    update_on_launch: true
    update_cache_timeout: 0
    wait: true
```

The following roles take the variable input to create inventories, projects, and inventory sources. This is an excerpt from the following file:

```
//inventory_creation/aap_inventory_role.yml
  roles:
    - redhat_cop.controller_configuration.projects
    - redhat_cop.controller_configuration.inventories
    - redhat_cop.controller_configuration.inventory_sources
    - redhat_cop.controller_configuration.
          inventory_source_update
```

Each different method of populating inventory sources has its use. The GUI is good for quick changes, modules are useful for making changes during jobs or playbooks, and roles are used for maintaining configuration. The next section goes into detail about the different types of inventory plugins that can be used in an inventory source.

Using base Ansible inventory plugins

Inventory plugins are Python scripts that can be utilized by both the `ansible-navigator` Ansible command line and the Automation controller. This section will cover existing popular inventory plugins. There are a few inventory sources that are built into the Automation controller that do not work like other plugins, such as the AWS inventory source. These will be discussed in the following *Using built-in automation controller inventory plugins section.*

Plugins have the following sources:

- Pre-installed with Ansible

- Collections

- The `inventory_plugins` folder of a given Git project

Let's discuss them further.

Pre-installed Ansible plugins

There are a few plugins that come installed in `ansible-core`. This is the version of Ansible that does not contain any additional collections.

The following command can be used to list all currently installed plugins that are available to be used by Ansible and from installed collections:

```
ansible-doc -t inventory -l
```

The built-in plugins fall into the following categories:

- String interpreters

- File readers

- Plugin and script readers

We'll now discuss each of them further.

String interpreters

There are two string interpreter plugins: `host_list` and `advanced_host_list`. These plugins are used in playbooks to interpret host and group lists. The next excerpt provides some examples of host lists being used; these can be found in the chapter code repository (`inventory_plugin/ inventory_host_list_examples.yml`).

At its base, this is a comma-separated list of hosts to use, as shown here:

```
- name: Example host list
  hosts: 'host1.example.com, host2'
  tasks:
```

However, the `advanced_host_list` plugin allows for ranges of hosts to be supplied as well. In this case, this would be `host1, host2, host3, ..., host10`:

```
hosts: 'host[1:10],'
```

Finally, as an example, this is not limited to just hostnames—it can be group names or even variables that contain comma-separated lists of hostnames, as shown here:

```
hosts: '{{ group_names }}, {{host_name}}'
...
```

These string interpreters are not used directly by inventory plugins in the Automation controller, but they are used to interpret every playbook that runs through either the Ansible **command-line interface** (**CLI**) or the Automation controller.

File readers

Ansible and the Automation controller can also interpret files that are in certain formats to extract hosts, groups, and variables. There are three built-in file interpreters: `ini`, `yaml`, and `toml`.

INI files

INI files are simple files. The variables can be after the host using a `key=value` pair. An excerpt of an example of this type of inventory file (`inventory_plugin/ini_inventory`) is shown next. The first example is a group with a host, and that host has a variable that follows it with the `key=value` pair:

```
[automationcontroller]
localhost ansible_connection=local
```

Variables can be denoted for an entire group by using a `:vars` modifier on a section, as illustrated in the following code snippet:

```
[automationcontroller:vars]
peers=execution_nodes
```

Child groups can be denoted by using the `:children` modifier, as illustrated in the following snippet. These hosts are also part of the parent group. A child group's variables have precedence over parent groups:

```
[automationcontroller:children]
sso
```

This is the child group:

```
[sso]
random_host_1
```

The `all` group contains all hosts automatically. Variables can be added to the `all` group just as they can to any other group, as illustrated in the following code snippet:

```
[all:vars]
admin_password=''
```

This second simple example demonstrates how to use an ungrouped host and a grouped host. Anything outside of an INI section is ungrouped:

```
# Example 2
ungrouped_host  # Ungrouped host

[grouped_host]
grouped_host1
```

A full list of rules around INI inventories can be found here:

`https://docs.ansible.com/ansible/latest/collections/ansible/builtin/ini_inventory.html`

YAML files

YAML is also an option for inventories. It uses nested dictionaries to set the values of hosts and groups.

A YAML inventory should start with the `all` group, as illustrated in the following code snippet, and have subsets of hosts and groups entities afterward:

```
//inventory_plugin/yaml_inventory.yml
all:
```

Each section expects a `hosts` and `vars` definition; however, this is not needed. Hosts don't have to belong to groups. Use indentation to move to a subsection. You can see an illustration of this in the following code snippet:

```
hosts:
    test1:
    test2:
        host_var: value
vars:
    group_all_var: value
```

Children denote a new section from which to make definitions. From here, the next section will denote the group name—in the following case, other_group. Subsections follow the same rules as the parent:

```
children:
    other_group:
        children:
            group_x:
                hosts:
                    test5
        vars:
            g2_var2: value3
...
```

A full list of rules around YAML inventories can be found here:

```
https://docs.ansible.com/ansible/latest/collections/ansible/
builtin/yaml_inventory.html
```

TOML files

Tom's Obvious Minimal Language (**TOML**) is a markup standard that can also be used to load inventories. It is similar to INI in some regards but has its own rules.

Each group is a table. The table can have parts such as hosts or vars. Variables are set in key=value pairs, as illustrated in the following code snippet:

```
//inventory_plugin/toml_inventory.toml
---
[all.vars]
some_var = false
```

Children and subsets are set in arrays, as shown here:

```
[api]
children = [
    "database"
]
```

Inline tables can also be used with { } deliminators, as illustrated here:

```
vars = { db_port = 8888, other_var = 'string' }
```

Child variables will take precedence over parent variables, as illustrated here:

```
[database.vars]
some_var = true

...
```

A full list of rules around TOML inventories can be found here:

https://docs.ansible.com/ansible/latest/collections/ansible/
builtin/toml_inventory.html

Plugin and script readers

There are other plugins as well that do not rely on files to load. Each of these serves a purpose. However, beyond `auto`, they are rarely used.

auto

The `auto` plugin is used for loading a Python script plugin from a YAML configuration file. These will be discussed in the *Using other popular inventory plugins* section.

constructed

The `constructed` plugin uses a Jinja2 construct using variables that are either from an existing inventory or variables already loaded into Ansible. A good writeup of a generated inventory can be found here: `https://cloudautomation.pharriso.co.uk/post/ansible-constructed-inventory-plugin/`.

generator

`generator` uses variables and patterns with a Jinja template to form a generated inventory. A writeup of how to use this with good visualization can be found here: `https://willthames.github.io/2017/11/01/generating-inventory.html`.

script

This is a Python inventory script that must accept the parameters of `--list` and `--host` that returns a **JavaScript Object Notation** (**JSON**)-formatted inventory. This has fallen out of use and was removed from the Automation controller in favor of inventory plugins.

With an overview of built-in Ansible plugins that can be used, there are more inventory plugins built into the Automation controller for use. Using those plugins will be the focus of the next section.

Using built-in Automation controller inventory plugins

There are several plugins built directly into the controller—these are mainly for connecting to cloud services. A list of supported services that can gather information about hosts, followed by their API and module aliases, is provided here:

- AWS **Elastic Compute Cloud (EC2)**—`ec2`
- **Google Compute Engine (GCE)**—`gce`
- Microsoft **Azure Resource Manager (ARM)**—`azure_rm`
- VMware vCenter—`vmware`
- Red Hat Satellite 6—`satellite6`
- Red Hat Insights—`insights`
- OpenStack—`openstack`
- **Red Hat Virtualization (RHV)**—`rhv`
- Red Hat Ansible Automation Platform—`controller`
- Project—`scm` will be covered in the following *Using other popular inventory plugins section*

Most of these inventory sources share a common form. Previously, the *Using inventory sources* section went into detail of how to add these as inventory sources to the Automation controller. There are also additional attributes that can be set for all inventory sources, such as host filters. The next subsections will go into detail about using each.

Each built-in inventory source shares two parts, as outlined here:

- A credential of a type that corresponds to the source, such as AWS or Insights
- Variables that need to be set for identifying hosts and groups to add to the inventory

In addition, all inventory plugins can have the following options:

- `keyed_groups`: Add hosts to groups for each `aws_ec2` host's `tags.Name` variable. `key` is the key to use, `prefix` is the prefix to ignore and not include in the group name, and `separator` is the divider for a list such as a comma or an underscore (_).

  For example, here's how to create groups for each AWS region:

  ```
  - key: placement.region   # aws_region_us_east_2
    prefix: aws_region
    separator: ""
  ```

This will put each host into a group based on the region. Using the preceding example, the group name would be us_east_2.

- groups: Add hosts to a group based on key:value pairs. The values and keys are dependent on the source.

  For example, here's how to add hosts to the 'east' group if any of the tags are east:

  ```
  groups:
    east: "'east' in (tags|list)"
  ```

- compose: Create host variables from key:value pairs.

  Here are some examples of this:

  ```
  ec2_state: state.name
  ec2_security_group_ids: security_groups |
  map(attribute='group_id') | list |  join(',')
  effectiverole: effectiveRole
  ```

- use_contrib_script_compatible_sanitization is a Boolean that defaults to false. This turns off the default group name sanitization. Review the documentation of each plugin to learn how this is applied. This sanitizes group names to be Ansible-safe. By using this, the TRANSFORM_INVALID_GROUP_CHARS Ansible configuration setting should be turned off. Use this option if there are problems with group names and other errors.

A good list of composed variables for the various supported plugins is on the documentation page for Automation controller inventory plugins, at the following link:

```
https://docs.ansible.com/automation-controller/latest/html/
userguide/inventory_plugins_templates.html
```

AWS EC2

The EC2 plugin connects to AWS to get hosts and populate an inventory. This can be narrowed down and customized. It is recommended to check out both of the documentation pages to get ideas on customizing your inventory.

The credential this source uses is **Amazon Web Services**.

This plugin has several variables to use, as outlined here:

- regions: A list of regions to use, such as those shown here:

  ```
  regions:
    - us-east-1
    - us-east-2
  ```

- `filters`: A list of filters to use, such as all instances with the environment tag set to `dev`, as illustrated here:

  ```
  filters:
    tag:Environment: dev
  ```

- `strict_permissions`: A Boolean that allows 403 errors to be skipped.

- Here are some other variable examples to use:

  ```
  compose:
    ansible_host: public_ip_address
  keyed_groups:
    - key: instance_type | regex_replace("[^A-Za-z0-9\_]",
  "_")
        parent_group: types
        prefix: type
  ```

The documentation page for this plugin can be found here:

`https://docs.ansible.com/ansible/latest/collections/amazon/aws/aws_ec2_inventory.html`

GCE

The Google plugin connects to Google Cloud to get hosts and populate an inventory.

The credential this source uses is **Google Compute Engine**.

This plugin has several variables to use, as outlined here:

- `projects`: A list of projects to use, as illustrated here:

  ```
  projects:
    - gcp-prod-gke-100
    - gcp-cicd-101
  ```

- `zones`: A list of zones to use, as illustrated here:

  ```
  zones:
    - us-east1-a
  ```

- `filters`: A list of filters to use, such as those of a particular machine type, as illustrated in the following code snippet. Filters in a list will be added as an AND to each other (filter1 and filter2), so the host must match all criteria:

```
filters:
  - machineType = n1-standard-1
```

- Other variable examples to use are shown here:

```
compose:
  ansible_ssh_host: networkInterfaces[0].
accessConfigs[0].natIP | default(networkInterfaces[0].
networkIP)

  gce_description: description if description else None

keyed_groups:
- key: gce_subnetwork
  prefix: network
```

The documentation page for this plugin can be found here:

```
https://docs.ansible.com/ansible/latest/collections/google/cloud/
gcp_compute_inventory.html
```

Microsoft ARM

Connects to Azure to search for hosts to add to an inventory.

The credential this source uses is **Microsoft Azure Resource Manager**.

This plugin has several variables to use, as outlined here:

- `exclude_host_filters`: A list of filters to use to exclude hosts. Operates on an OR expression, once a host that matches it excludes it. You can see an example usage of this here:

```
exclude_host_filters:
  - location in ['eastus']
  - powerstate != 'running'
```

- `hostvar_expressions`: `hostvar` names to map to Jinja2 values, similar to `compose`. You can see an example usage of this here:

```
hostvar_expressions:
  ansible_host: (public_dns_hostnames + public_ipv4_
addresses) | first
```

- `conditional_groups`: A mapping of group names to Jinja2 expressions. When the mapped expression is `true`, the host is added to the named group. Operates on an `OR` expression. You can see an example usage of this here:

```
conditional_groups:
    all_the_hosts: true
    db_hosts: "'dbserver' in name"
```

- `include_vm_resource_groups`: A list of resource group names to search for hosts in. A separate variable, `include_vmss_resource_groups`, supports searching scale sets. You can see an example usage of this here:

```
include_vm_resource_groups:
  - myrg1
  - myrg2
```

- Other variable examples to use are shown here:

```
keyed_groups:
  - prefix: tag
    key: tags
  - prefix: azure_loc
    key: location
```

The documentation page for this plugin can be found here:

```
https://docs.ansible.com/ansible/latest/collections/azure/
azcollection/azure_rm_inventory.html
```

VMware vCenter

Connect to vCenter to get hosts and groups to populate an inventory.

The credential this source uses is **VMware vCenter**.

This plugin has several variables to use, as outlined here:

- `strict`: A Boolean that defaults to `false`. It's possible to make expressions that are invalid; if `true`, it will fail if any expressions fail.

- `properties`: Uses VMware schema properties associated with the VM to populate `hostvars` in the inventory. You can see an example usage of this here:

    ```
    properties:
       - availableField
       - configIssue
    ```

- `with_nested_properties`: A Boolean that defaults to `true`. Transforms flattened properties' names to a nested dictionary.

- `filters`: Filters hosts based on Jinja2 templating. You can see an example usage of this here:

    ```
    filters:
       - runtime.powerState == "poweredOn"
    ```

- Other variable examples to use are shown here:

    ```
    compose:
       ansible_host: guest.ipAddress
       ansible_ssh_host: guest.ipAddress
       effectiverole: effectiveRole
    keyed_groups:
    - key: config.guestId
      prefix: ''
      separator: ''
    - key: tag_category.OS
      prefix: "vmware_tag_os_category_"
      separator: ""
    ```

The documentation page for this plugin can be found here:

`https://docs.ansible.com/ansible/latest/collections/community/vmware/vmware_vm_inventory_inventory.html`

Red Hat Satellite 6

The Ansible Automation Platform plugin allows another Automation controller or inventory on the same controller to act as a source for hosts.

The credential this source uses is **Red Hat Satellite 6**.

This plugin has several variables to use, as outlined here:

- `want*`: A Boolean—there are many `want` variables, and most are `true` by default. They determine if things such as **Internet Protocol 4** (**IPv4**), subnets, and other options are returned. The following are off by default: `want_facts`, `want_hostcollections`, and `want_params`. Look at the documentation link provided next for a full rundown of want variables.

- `group_prefix`: Prefix to apply to groups.

- `host_filters`: Restricts the list of returned hosts. You can see an example usage of this here:

      host_filters: 'organization="Web Engineering"'

- Other variable examples to use are shown here:

      keyed_groups:
      - key: foreman['location_name'] | lower | regex_replace('
      ', '') | regex_replace('[^A-Za-z0-9_]', '_')
        prefix: foreman_location_
        separator: ''

The documentation page for this plugin can be found here:

```
https://docs.ansible.com/ansible/latest/collections/theforeman/
foreman/foreman_inventory.html
```

Red Hat Insights

The Red Hat Insights variable connects to Red Hat Insights to gather hosts that are monitored by this service. This can be paired with a project and templates to use remediation on hosts that are offered by the Insights service.

The credential this source uses is **Insights**.

There are no variables needed to configure the Insights inventory.

The documentation page for this plugin can be found here:

```
https://docs.ansible.com/automation-controller/latest/html/
userguide/inventories.html#ug-source-insights
```

More information about an creating an Insights project can be found here:

```
https://docs.ansible.com/ansible-tower/latest/html/userguide/
insights.html
```

OpenStack

The Ansible Automation Platform plugin allows another automation controller or inventory on the same controller to act as a source for hosts.

The credential this source uses is **OpenStack**.

This plugin has several variables to use, as outlined here:

- `expand_hostvars`: A Boolean that defaults to `False` that can run extra commands to get additional information about hosts. This can be expensive in terms of time and resources.
- `groups`: Add hosts to groups based on Jinja2 conditionals.

The documentation page for this plugin can be found here:

```
https://docs.ansible.com/ansible/latest/collections/openstack/
cloud/openstack_inventory.html
```

RHV

The RHV plugin gets information about hosts managed by RHV.

The credential this source uses is **Red Hat Virtualization**.

This plugin has several variables to use, as outlined here:

- `ovirt_insecure`: This Boolean variable will determine whether **Transport Layer Security (TLS)** certificates will be checked when connecting to the host. There is no default option.
- `ovirt_query_filter`: This filters key-values when querying VMs. The documentation page has more fields to use for this filter.

  Here's an example usage of this:

  ```
  search: 'name=myvm AND cluster=mycluster'
  case_sensitive: no
  max: 15
  ```

- `ovirt_hostname_preference ["fqdn", "name" ]`: A list of the order of preference in which hostnames are assigned. The documentation page has more attributes that can be used.
- Some variable examples to use are shown here:

  ```
  compose:
    ansible_host: (devices.values() | list)[0][0] if
  devices else None
  ```

```
    keyed_groups:
    - key: cluster
      prefix: cluster
      separator: _
```

The documentation page for this plugin can be found here:

`https://docs.ansible.com/ansible/latest/collections/ovirt/ovirt/ ovirt_inventory.html`

Red Hat Ansible Automation Platform

The Ansible Automation Platform plugin allows another automation controller or inventory on the same controller to act as a source for hosts.

The credential this source uses is **Red Hat Ansible Automation Platform**.

The variable that needs to be set is the inventory **identifier** (**ID**) of the inventory to reference.

The inventory ID can be found from the GUI by referencing the **HyperText Markup Language** (**HTML**) address when on that inventory. For example, `https://controller.node/#/ inventories/inventory/2/details` would be inventory with an ID of 2.

The other option is to search for the inventory ID with an API request. This request would use the form `https://controller.node/api/v2/inventories/?name=localhost`, where the name `localhost` is the name of the inventory.

The documentation page for this plugin can be found here:

`https://docs.ansible.com/ansible/latest/collections/awx/awx/ controller_inventory.html`

This section covered built-in inventory plugins in the Automation controller. However, the controller is not limited to only these plugins—additional plugins can be used when sourced from a project. The next section will go into detail about how to use these plugins.

Using other popular inventory plugins

So far, the inventory plugins and sources that have been covered have been those built into Ansible and the Automation controller. There are other sources out there for inventory plugins that can be used. Two such examples are NetBox and ServiceNow. They are loaded differently than built-in plugins. The next section goes into the details of loading custom inventory plugins.

As they need a collection installed from a project and a definition file, they can take advantage of custom credential types to populate sensitive and common data. These are not the only two sources, but they are essentially custom inventory plugins included in a collection.

Overview of using custom inventory plugins

Using custom inventory plugins are sourced from `scm/git` projects. The creation of projects in the Automation controller will be covered in *Chapter 10, Creating Job Templates and Workflows*. This section will go into detail about how the project folder should be structured.

At the top of the project should be a `collections` folder with a `requirements` file, and then there should be a file to reference for each inventory source that is used by that collection. An example of this is shown here:

```
├── collections
│   └── requirements.yml
└── netbox.yml
```

The `requirements` file is a simple file that the Automation controller will use to load the collection from Automation hub or Ansible Galaxy. It should look like this:

```
---
collections:
- name: netbox.netbox
...
```

The last piece of the puzzle is the definition file. Some plugins require the name to have a certain form; for example, the ServiceNow plugin requires the file to end in either `now.yml` or `now.yaml`. These files require a variable to be set for the plugin to set which plugin to use to read the file. It also has the option for `compose` and `keyed_groups`, just as the built-in plugins do. The plugin documentation gives more information about what is required for each service.

With the basics set, the next section dives into the details of using these inventory plugins.

NetBox

NetBox is a popular SOT for network devices. Its collection name is `netbox.netbox`. A full example of a NetBox inventory can be found in this chapter's repository at `netbox/netbox.yml`. In addition, there is a `netbox/netbox_creds.yml` credentials file that can be used to create a credential type and credential for authenticating to NetBox.

The file should start with the plugin to use, as shown here:

```
plugin: netbox.netbox.nb_inventory
```

There are several authentication credential variables to use. These can be set with an Automation controller credential/credential type with the `netbox/netbox_creds.yml` file provided in this chapter's repository, or just be included in the definition file. Here are a couple of examples:

- `api_endpoint`: Endpoint of the NetBox API
- `token`: API token used to be able to read NetBox

Then, there are several Boolean variables to use. This list includes a description of them and their default value:

- `validate_certs`: `True`; whether or not to verify SSL of the host.
- `config_context`: `False`; add configuration context to `hostvars`. Check out the NetBox documentation for more context.
- `services`: `True`; adds device services information to `hostvars`.
- `interfaces`: `False`; adds device interface information to `hostvars`.
- `fetch_all`: `True`; whether to retrieve all information about the device. If using query filters, this may make the plugin unusable and may need to be turned off.

In addition, there are other variables to use to filter out hosts and group hosts, as outlined here:

- `query_filters`: A list of parameters passed to the query string for both devices and VMs

  Here's an example of this:

  ```
  query_filters:
      - cf_foo: bar
      - status: active
      - tag: tag_name
  ```

- `group_by`: Keys used to create groups

  Here's an example of this:

  ```
  group_by:
      - device_roles
      - racks
  ```

- Here are some variable examples to use:

  ```
  compose:
      foo: last_updated
      bar: display_name
  ```

The documentation page for this plugin can be found here:

```
https://docs.ansible.com/ansible/latest/collections/netbox/netbox/
nb_inventory_inventory.html
```

ServiceNow

ServiceNow is a popular SOT for VMs. Its collection name is `servicenow.servicenow`. A full example of a ServiceNow inventory can be found in this chapter's repository at `servicenow/service_now.yml`. In addition, there is a `servicenow/servicenow_creds.yml` credentials file that can be used to create a credential type and credential for authenticating to ServiceNow.

The file should start with the plugin to use, as illustrated here:

```
plugin: servicenow.servicenow.now
```

There are several authentication credentials to use. These can be set with an Automation controller credential/credential type with the file provided in this chapter's repository or just be included in the definition file. Some of them are listed here:

- `host`: The ServiceNow **fully qualified domain name (FQDN)** hostname. Exclusive with the instance.

- `instance`: ServiceNow instance name without the domain. Exclusive with the host.

- `username`: Name of the user to use for authentication.

- `password`: Password to use for authentication.

- `table`: The ServiceNow table to query. You can see an example usage of this here:

    ```
    table: cmdb_ci_netgear
    ```

- `fields`: Additional table columns to add to the `hostvars` of each host, as illustrated in the following example:

    ```
    fields:
      - name
      - sys_tags
    ```

- Here are some variable examples to use:

    ```
    compose:
      sn_tags: sn_sys_tags.replace(" ", "").split(',')
      ansible_host: sn_ip_address
    keyed_groups:
    ```

```
    - key: sn_classification | lower
      prefix: 'env'
    - key: sn_vendor | lower
      prefix: ''
      separator: ''
```

The documentation page for this plugin can be found here:

```
https://docs.ansible.com/ansible/latest/collections/servicenow/
servicenow/now_inventory.html
```

With an overview of built-in inventory plugins and custom plugins that have been prebuilt, the next step is to create your own inventory plugin.

Writing your own inventory plugin

While the inventory plugins that have been covered so far cover the most popular services, it is impossible for them to cover every possibility. A good example of this is a SOT that has been custom-built for a team or company. The easiest way to access something like this is through an API. However, it does not have to use an API and can instead make use of Python to access data to build an inventory.

Ansible configuration changes

As discussed, when using collection plugins, a project should have a certain structure to it. In addition, there are advantages to testing to structure it in a particular order. The recommended structure to use is that of the `inventory_plugin_creation` folder in this chapter's repository files, as illustrated here:

```
├── ansible.cfg
├── inventories
│   └── plugin.yml
├── inventory_plugins
│   └── publicapis.py
└── readme.md
```

In this example, the base directory contains a `readme.md` file and an `ansible.cfg` file. The `inventories` folder contains a `plugin.yml` plugin definition file, and the `inventory_plugins` folder contains a `publicapis.py` plugin script.

The default Ansible configuration for where to store inventory plugins is deep inside the Ansible libraries, or in a user folder, which makes the default location impractical for use in the Automation controller. To change where the inventory plugin is pulled from, a change needs to be made to the Ansible configuration. In addition, the inventory plugin itself must be whitelisted in order to be allowed to run. To do this, make the following entries in the `ansible.cfg` file:

```
//inventory_plugin_creation/ansible.cfg
---
[defaults]
inventory_plugins = ./
[inventory]
enable_plugins = publicapis, auto
```

This sets the base folder to look for in the current directory and whitelists the `publicapis` plugin. Multiple plugins can be placed in the `inventory_plugins` folder. If for some reason they should reside in multiple folders, the way to allow that is to use a colon delimitator in the `folder1:folder2` value, as for the `inventory_plugins` setting.

Custom inventory plugin script

For the purposes of this section, an inventory plugin has been created that uses the public API at `https://api.publicapis.org/entries`. The next few sections will reference excerpts from the following file in this chapter's repository:

```
//inventory_plugin_creation/inventory_plugins/publicapis.py
---
DOCUMENTATION = r"""
    inventory: publicapis
    name: Public APIs
    plugin_type: inventory
```

The first part of an inventory plugin is the documentation. This is important as the options from the documentation are checked as inputs. The following fields are used by the documentation portion of the plugin:

- `inventory`: Name of the plugin. This must match the filename of the plugin Python script, and the plugin noted in the definition file.
- `plugin_type`: Type of plugin—always `inventory` in this case.
- `author`: A list of authors of the plugin. This can also include contact information.

- short_description: A short description of the plugin, which shows up near the top on the documentation page for Galaxy or Automation hub.

- version_added: The version of Ansible this was initially added to.

- description: A list with a longer description of the plugin.

- options: Options for input from the definitions file. This is discussed in the *Plugin options* subsection next.

- requirements: A list of requirements to run the add-on; for example, the version of Python required, such as python >= 3.4.

Plugin options

The plugin options section contains details of the inputs the plugin will accept. This can include usernames, passwords, hosts, or other options. The options take on the following form:

```
option_name:
    description: description
    type: string
    aliases: [ option_alias ]
    required: True
    env:
        - name: ENV_OPTION
```

For each input option for the plugin shown previously, the following variables are used to describe the option:

- option_name: Name of the option; any valid Ansible variable.

- description: A description of the option.

- type: *STRICTLY* enforced type of variable; this is either string, list, or bool.

- aliases: A list of aliases for the option.

- required: Boolean, if the option is required.

- env: The name of the environment variable that can be used instead of the definition file or variable input.

Plugin examples

The next section is the examples section. There are very few rules around this section. It should be in a markdown format. Here's an example of what this section might look like:

```
EXAMPLES = r"""
plugin: publicapis
validate_certs: False
publicapi_url: 'https://api.publicapis.org/entries'
"""
```

Plugin code

This section will handle the actual Python code to create an inventory plugin in the plugin file.

Imports

It is important to import the correct Ansible Python libraries to facilitate the inventory plugin. The base imports are shown here:

```
# Ansible internal request utilities
from ansible.module_utils.urls import Request, ConnectionError, urllib_error
from ansible.errors import AnsibleError, AnsibleParserError
from ansible.plugins.inventory import BaseInventoryPlugin, Constructable
```

These import useful Python functions from Ansible, the most important being `BaseInventoryPlugin`, which is used to create inventories.

In this section of code, any other imports that are needed should be declared. For example, to import `os` and `requests`, execute the following code:

```
import os
try:
    import requests
except ImportError:
    raise AnsibleError("Python requests module is required for this plugin.")
```

These are useful for dealing with file reading and making API requests.

Class invocation

The class invocation is next. This creates the `inventory` module and sets the name of the inventory. The name must match the inventory name from the documentation, which should also be the same as the plugin invoked in the definition file. The following code snippet provides an example of this:

```
class InventoryModule(BaseInventoryPlugin, Constructable):
    NAME = "publicapis"
```

Initialization

It is important to initialize `InventoryModule` and the variables that will be used. This is done with an `init` command and by declaring global variables used. Here's an example of this:

```
def __init__(self):
    super(InventoryModule, self).__init__()

    # from config
    self.validate_certs = None
    self.publicapi_url = None
    self.session = None

    self._inventory = None
    self._host_list = None
```

File verification

An important inclusion in the inventory module is file verification, as this helps make sure it picks up the right definition file. An example of this section is shown here:

```
def verify_file(self, path):
    valid = False
    if super(InventoryModule, self).verify_file(path):
        if path.endswith(("plugin.yaml", "plugin.yml")):
            valid = True
        else:
            self.display.vvv(
                'Skipping due to inventory source not
ending in "plugin.yaml" nor "plugin.yml"'
            )
    return valid
```

This section checks that the path does exist and that it is named correctly. If it does not find a file that matches or a file doesn't exist, then the plugin will be skipped. This is important when using installations or references that have multiple inventory plugins.

Parsing section

The `parse` function is the main section of the inventory plugin. In this section, it reads in the `definitions` file and the options and initializes the root group, which should be `group_all`. The following code snippet provides an illustration of this:

```
def parse(self, inventory, loader, path, cache=True):
    super(InventoryModule, self).parse(inventory, loader, path)
    self._read_config_data(path)

    # Get inputs.
    self.validate_certs = self.get_option("validate_certs")
    self.publicapi_url = self.get_option("publicapi_url")

    # Create root group
    root_group_name = self.inventory.add_group("group_all")

    self.query_entries(root_group_name)
```

At the end of this section, we pass the root group to a query function that is built to parse through the API.

Looping over API results

The next section of code gets the results from the API and initiates a `for` loop. This section of code can be modified to work with any API and is built as such, as illustrated here:

```
        query = self.publicapi_url
        r = requests.get(query, timeout=10, verify=self.
validate_certs)
        query_result = r.json()
        entries = query_result["entries"]
        for entry in entries:
```

After the request is made, the hostname is set. The group name is set, and there is a check that the group name does not equal the hostname. There is an issue that arises when adding children if the hostname and group name match, so a workaround is to modify the group name, as follows:

```
host_name = self.inventory.add_host(entry["API"])
if host_name == entry["Category"]:
    group_name = entry["Category"] + '_group'
    self.inventory.add_group(group_name)
else:
    group_name = self.inventory.add_group(entry["Category"])
```

The final section is for populating the variables for hosts, as illustrated here:

```
for field_name in (
    "Description",
    "Auth",
):
    self.inventory.set_variable(host_name, field_name,
entry[field_name])
```

The set_variable function can be used to add variables to hosts and groups.

There are a few more functions that were not covered in the examples, as the API did not lend itself to their use.

Additional Python functions

The to_safe_group_name is a way to sanitize group names and customize how unsafe characters are replaced. The following code snippet shows the import and function:

```
from ansible.inventory.group import to_safe_group_name
to_safe_group_name(name, replacer='')
```

Constructable can be imported to use keys such as the compose and keyed group functions, as referenced in the built-in plugins. There is also a document fragment to extend for Ansible and Automation hub documentation, as illustrated here:

```
from ansible.plugins.inventory import BaseInventoryPlugin,
Constructable
extends_documentation_fragment:
  - constructed
```

The add_host_to_composed_groups function does as it says, along with the add_host_to_keyed groups and set_composite_vars. Using the format from the built-in plugins, this is what it looks like:

```
self._add_host_to_composed_groups(self.get_option('groups'),
host_vars, hostname, strict=strict) self._add_host_to_keyed_
groups(self.get_option('keyed_groups'), host_vars, hostname,
strict=strict)
self._set_composite_vars(self.get_option('compose'), host_vars,
hostname, strict=True)
```

These functions rely on the composition of hostvars being set and the documentation fragment being added so that the option can be used in the definitions file.

This section went into the details of creating a custom inventory plugin. These tools provide a framework for adapting any source of data about hosts to be easily accessible in the Automation controller.

Summary

While a lot of options have been presented, it is best to use a plugin that matches where the host is stored. If you are using physical machines, ServiceNow or a custom plugin may be appropriate, whereas if you are using a cloud provider, their plugin is likely the best to use. The plugins allow you to tailor inventories and sources as needed.

This chapter covered inventories, inventory plugins, and inventory sources, which are pivotal to running jobs in the Automation controller. The other requirement for job templates is EEs.

The next chapter will cover the creation of custom EEs for use in job templates.

Further reading

For more reading about the use and making of custom inventory plugins, the following resources are recommended:

- Developing inventory documentation: `https://docs.ansible.com/ansible/latest/dev_guide/developing_inventory.html`

- Managing meaningful inventories: `https://github.com/willtome/managing-meaningful-inventories`

- Inventory file examples: `https://github.com/AlanCoding/Ansible-inventory-file-examples`

8

Creating Execution Environments

Execution environments (**EEs**) replace virtual environments in terms of where playbooks and other jobs are run in the Automation controller. They are containers that contain system dependencies, Python libraries, a version of Ansible, and Ansible collections. This chapter will describe what EEs are and what goes into them, their benefits, and how they can be used in both the Automation controller and the command line. You will also learn how to create and manage EEs using Ansible Builder. Finally, you will learn how to use a role to build an EE and push it to Automation hub.

In this chapter, we will cover the following topics:

- What are execution environments?
- How to use execution environments
- Creating and modifying your execution environments
- Using roles to create execution environments

Technical requirements

All the code referenced in this chapter is available at `https://github.com/PacktPublishing/Demystifying-Ansible-Automation-Platform/tree/main/ch08`. It is assumed that you have Ansible installed to run the code provided.

This chapter assumes that you have already installed Python and Ansible.

The software for this chapter can be installed with the following commands:

- `podman`: For managing containers:

  ```
  $ sudo dnf -y install podman
  ```

- `ansible-builder`: Used to build EEs:

  ```
  $ pip install ansible-builder
  ```

- `ansible-navigator`: Used to execute playbooks in EEs:

  ```
  $ pip install ansible-navigator
  ```

- `podman login`: Used to log in to container registries:

  ```
  $ podman login <registery_fqdn>
  ```

In terms of `podman login`, you must enter a username and password. This allows you to authenticate to a credentialed registry so that containers can be both pulled from and pushed to it.

What are execution environments?

Under the previous major version of Ansible Automation Platform, jobs and playbooks were run with bubblewrap to isolate the process. They also took advantage of Python virtual environments. However, managing these virtual environments and module dependencies was challenging and labor-intensive. EEs are the answer to this.

EEs are prebuilt containers that are made to run Ansible playbooks. These replace using Python virtual environments as the standard way of using different versions of Python and Python packages. Ansible's Automation controller takes advantage of these environments to scale and run job templates as well. They solve the issue of, *it works for me*, maintaining different environments across all nodes, and other problems that arose from the previous solution. They also double as a simplified developmental version of the Automation controller for testing a job template when using the same container as the controller.

What is inside the execution environment?

The execution environment is made up of a few things, as shown in the following diagram:

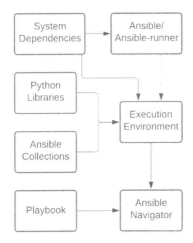

Figure 8.1 – Execution environment pieces

An execution environment contains the following pieces:

- **Ansible**: An installation of ansible-core – that is, the base version of Ansible without collections.

- **Ansible Runner**: A tool used to invoke Ansible and consume its results.

- **System Dependencies**: System dependencies installed by dnf.

- **Python Libraries**: Python libraries installed with pip.

- **Ansible Collections**: Collections installed from a requirements file.

The execution environment container image is then spun up as a container by either the Automation controller or Ansible Navigator. After this, the playbook or project directory and other pertinent information are copied over and Ansible is run.

Seeing what is in a specific execution environment

Each EE can be different, and it can be difficult to tell what pieces are inside them at a glance. The ansible-navigator tool has built-in functions that allow users to explore what each EE contains.

To get a particular image from Automation hub, follow these steps:

1. Navigate to Automation hub.
2. Click **Execution Environments**.
3. Select an execution environment from the list, such as ee-supported-rhel8.

4. Use the `podman pull` command from the on-screen instructions to make a local copy of the container:

    ```
    $ podman pull ah.node/ee-supported-rhel8:latest
    ```

    Ansible Navigator has a specific function that allows it to inspect container images currently on the local machine.

5. To explore a given EE image, run the `$ ansible-navigator images` command:

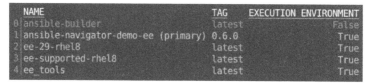

Figure 8.2 – ansible-navigator image list

6. Select an image from the list by pressing the number that relates to the respective image, as shown in the preceding screenshot.

7. Choose one of the following options to explore more of a given image:

    •   **Image information**: General information about the container image

    •   **General information**: Information about the operating system and Python version that is inside the image

    •   **Ansible version and collections**: The version of Ansible that's been installed, and the names and versions of the collections that have been installed

    •   **Python packages**: The names and versions of the Python packages that have been installed

    •   **Operating system packages**: The names and versions of the system packages that have been installed

    •   **Everything**: A YAML representation of every option mentioned previously

Now that you know how to look inside EEs, the next step is to use them.

How to use execution environments

Ansible Navigator allows an execution environment to be run from the command line. The following `demo.yml` playbook is available in the `ch01` folder:

```
//demo.yml
- - -
```

```
- name: Demo Playbook
  hosts: localhost
  gather_facts: false
  tasks:
    - debug:
        msg: Hello world
...
```

To run this playbook in an EE, use the `ansible-navigator run demo.yml -m stdout` command. It should output a `Hello world` message. Using the `-ee` or `-eei` option, the EE can be specified. This allows the user to use the same EE that was used in the controller for testing and development:

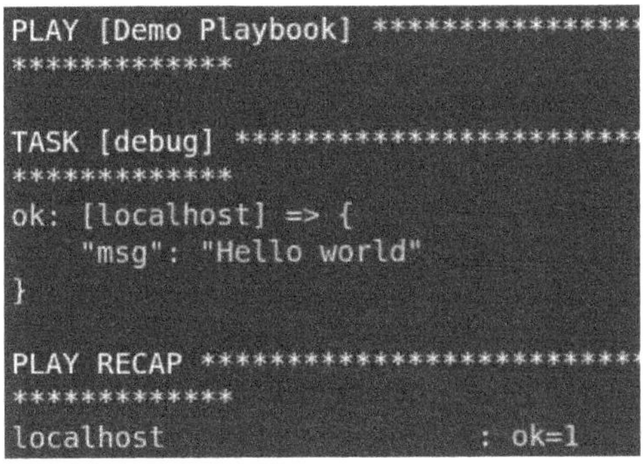

Figure 8.3 – ansible-navigator demo playbook

Inside the Automation controller, all jobs are run in an EE. When a playbook runs, a project or inventory is synced, or an ad hoc command is considered as a job.

Running a simple job in an execution environment

EEs are used for project updates and running jobs. To test an EE using ad hoc commands in the automation controller, follow these steps:

1. Navigate to the Automation controller.

2. Navigate to **Inventories**, choose an inventory, and select the **Host** tab.

3. Select a host using the checkboxes on the left-hand side.

4. Click **Run Command**, as shown in the following screenshot:

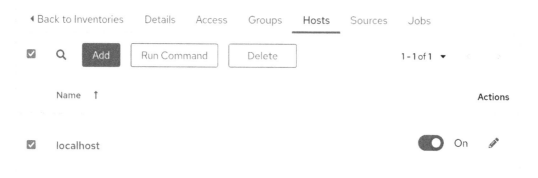

Figure 8.4 – Selecting a host for running an ad hoc command

5. Choose a module to use. It is recommended to use `setup` for demonstration purposes. Click **Next**.

6. Select an EE to use, such as `Automation Hub Default execution environment`. Click **Next**.

7. Choose a credential to use and click **Next**.

8. Click **Launch**. The output should look as follows:

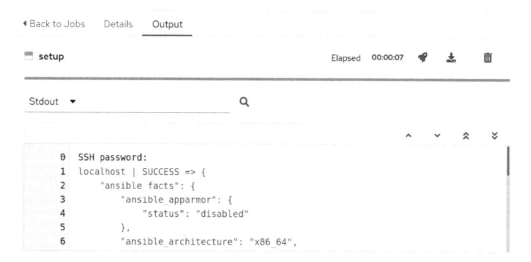

Figure 8.5 – Job output for the ad hoc setup command

Now that we've learned what can be found in EEs and how to use the default EEs, the next step is to create them.

Creating and modifying your execution environments

The EEs that come with Ansible Automation Platform can perform most general tasks, though some collections have different Python or system requirements that aren't included in a base container image.

To allow anyone to create an EE, the Ansible Builder tool was introduced. This tool takes a definition file and creates an EE.

Creating a definition file

The definition file is made up of four sections, as follows:

- **Base section**: This is where the base variables are set.
- **Build arguments section**: This is where the build arguments are set.
- **Dependencies section**: This is where definitions for dependencies are set.
- **Build steps section**: This is where additional build arguments are set.

To follow along, a base execution file has been created in this chapter's files:

```
//base/base_execution_enviroment.yml
---
version: 1

build_arg_defaults:
  EE_BASE_IMAGE: 'ah.node/ee-minimal-rhel8:latest'
  ANSIBLE_GALAXY_CLI_COLLECTION_OPTS: "-v"

dependencies:
  galaxy: requirements.yml
  python: requirements.txt
  system: bindep

additional_build_steps:
  prepend: |
    RUN whoami
    RUN cat /etc/os-release
  append:
    - RUN echo This is a post-install command!
```

This file is a good reference file and will be used in the upcoming examples. The first section to review is the base section.

Base section

The base section contains two parts:

- The version, which sets the version of the `ansible-builder` definition. At the time of writing, the only option is `1`. However, it is a good practice to keep it so that you know what version it is compatible with in the future:

  ```
  version: 1
  ```

- The second field, `ansible_config`, which provides the path to an Ansible configuration file. This is useful for connecting to a private Automation hub. An example file to connect to Automation hub can be found in this book's GitHub repository and is called `ch08/ansible.cfg`:

  ```
  ansible_config: 'ansible.cfg'
  ```

With the base section covered, the next step is the build arguments section.

The build arguments section

This section goes over the build arguments to use:

- The first variable is the base image to use. It is recommended to use the minimal EE file from Automation hub to build a new EE:

  ```
  build_arg_defaults:
    EE_BASE_IMAGE: 'ah.node/ee-minimal-rhel8:latest'
  ```

- The second variable specifies the `ansible-galaxy` options to use. This can be skipped, but if you're diagnosing problems with connecting to Automation hub or collections, it can be helpful to add the `-v` verbosity option. `-c` can also be used to set the verify SSL option to `false`. To invoke these options use the following variable:

  ```
  ANSIBLE_GALAXY_CLI_COLLECTION_OPTS: "-v"
  ```

- The last variable is the builder image to use. This is the image that is used for compiling tasks. It may be useful to set this behind a firewall and point the builder at another registry. If this is not an issue, just use the default and do not include this variable:

  ```
  EE_BUILDER_IMAGE: ansible-builder-rhel8:latest
  ```

With the build arguments section covered, the next section covers how to add dependencies to the EE.

Dependencies section

The next section is the dependencies section. It contains the files that the builder refers to so that they can be added to the final execution environment:

```
dependencies:
```

Galaxy is a standard galaxy requirements file that includes a list of roles and collections to install. Each of the collections can also contain galaxy, Python, and `bindep` requirements that have been added to the container:

```
galaxy: requirements.yml
python: requirements.txt
system: bindep.txt
```

Each of these is covered in more detail, the first being the galaxy dependencies.

Galaxy dependencies

The galaxy requirements file is in the following format and allows roles and collections to be added:

```
//base/requirements.yml
---
collections:
  - name: redhat_cop.ah_configuration
  - name: redhat_cop.controller_configuration
```

The collections can also supply their own Python dependencies. Any custom Python modules needed are covered in the next section.

Python dependencies

The Python requirements file is simple and based on the `pip freeze` command:

```
//base/requirements.txt
pytz  # for schedule_rrule lookup plugin
python-dateutil>=2.7.0  # schedule_rrule
awxkit  # For import and export modules
```

System dependencies

The system requirements must be installed using `yum` or `dnf`:

```
//base/bindep
python38-requests [platform:centos-8 platform:rhel-8]
python38-pyyaml [platform:centos-8 platform:rhel-8]
```

With dependencies covered the next section goes into the build step arguments that can be added to the container build process.

Build steps section

The last section is one for additional build steps. These are set to run before and after the execution environment is built:

```
//base/base_execution_enviroment.yml
additional_build_steps:
```

Both of these can take multi-line strings, as shown here:

```
prepend: |
   RUN whoami
   RUN cat /etc/hosts
```

They can also take a list of commands, as shown here:

```
append:
   - RUN echo This is a post-install command!
```

To add files to a container, place them in the same directory that's in the `context` folder. These can be added either with `prepend` or `append`. For example, you can use the following command to copy a Kerberos definition file from the `context` folder to the container:

```
   - COPY krb5.conf /etc/krb5.conf
```

With the definition file created, the next step is to use it to build the EE.

Creating the execution environment

The definition file is the blueprint for building the execution environment. The `ansible-builder` command takes that blueprint, grabs all the pieces, and puts things together. An important build argument is the container name. By default, it is `ansible-execution-env:latest`. The `-t` command will name the resulting container.

Use the following command to use builder:

```
$ ansible-builder build -f base_execution_enviroment.yml -t
base_ee
```

This will result in a container image being produced. To see the list of container images, run the following command:

```
$ podman image list
```

The image should be listed under `localhost/base_ee` if you named it with the previous command.

To push the new EE to Automation hub, follow these steps:

1. Retag the image with the remote registry in the name. You can also rename the image here:

    ```
    $ podman tag localhost/base_ee ah.node/base_ee
    ```

2. Push the new image to Automation hub:

    ```
    $ podman push --tls-verify=false ah.node/base_ee
    ```

These steps can be repeated and set to update EEs as needed, add more dependencies, or add anything additional to the custom EE as needed. The next section will explore creating EEs with roles and converting virtual Python environments into EEs.

Using roles to create execution environments

The command line is the way to use Ansible Builder, and there are roles to help build execution environments using **Configuration as Code** (**CaC**). This section will cover two roles from `redhat_cop.ee_utilities`. The first is the `ee_builder` role, which is built to take inputs and build out an execution environment. The second role is `virtualenv_migrate`, which is built to convert a Python virtual environment from a legacy Ansible Tower installation into an EE using the `ee_builder` role.

Creating an execution environment using the builder role

This role is useful for creating a CI/CD utility that can update execution environments when change is required. This is important as Red Hat releases regular updates for the base EE images, and it can also be used to add to an EE over time. Using CaC to define and create the EE makes managing it easy.

The `ee_builder` role takes the following variables. They are the variable equivalents of the sections from the execution definition file:

- `builder_dir`: The directory that will store all the build and context files
- `ee_registry_dest`: The path or URL where the image will be pushed
- `ee_registry_username`: The username to use when authenticating to remote registries
- `ee_registry_password`: The password to use when authenticating to remote registries
- `ee_image_push`: Bool; whether to push to the registry or not

The EE definitions are set in a list under the `ee_list` variable:

- `name`: Name of the EE image to create. This can include the tag for the container.
- `bindep`: The variable list to provide the `bindep` requirements if you're using variables
- `python`: The variable list to provide Python requirements.
- `collections`: The variable list to provide galaxy requirements if using variables, in `ansible-galaxy` list form.
- `prepend`: Additional build arguments in list form.
- `append`: Additional build arguments in list form.

The following excerpt shows how to use the role and variables that are included in this chapter's GitHub repository:

- These variables set where you can build, as well as the authentication to push the created EE to Automation hub:

  ```
  /roles/ee_builder_base.yml
  ---
    vars:
      builder_dir: /tmp/builder_base
      ee_registry_dest: ah.node/
      ee_list:
  ```

  Each item in the `ee_list` describes a single EE.

- `name` can contain a tag so that versioning can be applied:

  ```
  name: custom_ee:1.10
  ```

- `bindep` is the same as in the file, just in list form:

```
bindep:
    - python38-requests [platform:centos-8
platform:rhel-8]
    - python38-pyyaml [platform:centos-8
platform:rhel-8]
```

- `python` is a list and can contain versions:

```
python:
    - pytz  # for schedule_rrule lookup plugin
    - python-dateutil>=2.7.0  # schedule_rrule
    - awxkit  # For import and export modules
```

- `collections` can also take a version (this is optional):

```
ee_collections:
    - name: awx.awx
    - name: redhat_cop.controller_configuration
      version: 2.1.0
    - name: redhat_cop.ah_configuration
```

- The `prepend` and `append` options can create list forms if needed:

```
prepend:
    - RUN whoami
    - RUN cat /etc/os-release
append:
    - RUN echo This is a post-install command!
```

- `roles` can be invoked to create an EE and push it to Automation hub:

```
roles:
    - redhat_cop.ee_utilities.ee_builder
```

Using what's in this playbook, you can implement a CI/CD to adjust or update an EE over time.

You can also use the `ee_builder` role in tandem with the other role in the `ee_utils` collection to convert Python virtual environments into EEs.

Converting Python virtual environments from older Tower installations

Some places may still be using an older version of Ansible Automation Platform; that is, Tower. Tower uses Python virtual environments to manage Python modules. A role was created to facilitate migration from one to the other.

The `virtualenv_migrate` role takes the following variables. These are the variable equivalents of the sections in the execution definition file:

- `venv_migrate_default_ee_url`: The URL and container name of the EE to use as a base to compare the Python virtual environment.
- `ee_registry_username`: The username to use when authenticating to the remote registry.
- `ee_registry_password`: The password to use when authenticating to the remote registry.
- `CUSTOM_VENV_PATHS`: The Ansible Tower setting the role uses to look for virtual environments. This is set in Tower, not the role.

The role is built to be run as the host that Ansible has access to. This can be done with an inventory file. An inventory has been created in this chapter's GitHub repository that can be used for this purpose:

```
//roles/inventory.ini
---
[tower]
Tower.node

[tower:vars]
ansible_become=yes
ansible_user=excalibrax
ansible_become_password=password
```

A playbook has also been created in this chapter's GitHub repository that contains these two roles:

```
//roles/ee_venv_migrate.yml
```

For the role itself, only a simple variable needs to be set for the EE to compare it to:

```
    venv_migrate_default_ee_url: ah.node/ee-minimal-
rhel8:latest
```

Note that the variable specifies the registry, the EE name, and the tag to use.

With that set, the role can be invoked:

```
- name: Include venv_migrate role
  include_role:
    name: redhat_cop.ee_utilities.virtualenv_migrate
```

A second playbook in the same file has been set to target localhost. This is where the builder is invoked. Because it's the same role as in the previous section, they are the same inputs. However, it uses the Python dependencies from the virtual environments for migration purposes. It is assumed that additional collections would be added to this.

The following variables can be used:

```
venv_migrate_default_ee_url: ah.node/ee-minimal-
rhel8:latest
ee_registry_dest: ah.node/
ee_bindep: []
ee_collections:
  - name: awx.awx
  - name: redhat_cop.controller_configuration
  - name: redhat_cop.ah_configuration
```

To grab the variable from the first host in the Tower group, use the following code:

```
- set_fact:
    venv_migrate_ee_python: "{{ hostvars[groups['tower']
[0]]['venv_migrate_ee_python'] }}"
```

With all the variables set, the role can be invoked:

```
- name: Create EEs
  include_tasks: create_ee.yml
```

Migrating from the older version of Tower to EEs can be a pain, depending on the number of environments that were previously used. It can be useful to export this information as a file for future reference, or so that it can be converted into CI/CD format:

```
- name: Export python virtual enviroment list to file
  copy:
    content: "{{ venv_migrate_ee_python | to_nice_yaml(
width=50, explicit_start=True, explicit_end=True) }}"
    dest: venv_migrate_ee_python.yaml
```

Using these two roles should make migrating and creating execution environments easy.

Summary

In this chapter, we looked at EEs in detail, including what goes into them, how to define them, and the tools that can be used to create them. This book heavily pushes CaC for managing everything, and that cannot be more true for EEs. During the first half of 2022, Red Hat was publishing updates to the EEs for Ansible Automation Platform every few weeks. To keep up with security and bugfix updates, it is almost mandatory to put EE creation into an automated process to keep up with the changes.

The next chapter will focus on Automation hub. This is a repository for collections and EEs. Here, you will learn how to add these and manage Automation hub.

Automation Hub Management

The final thing you need to do as part of a job template is manage collections and **execution environments** (**EEs**). Automation hub acts as a repository for both collections and EEs. These are then consumed by the Automation controller.

In this chapter, we will provide an overview of the content that goes into Automation hub before learning how to manage collections. EEs and external registries will also be covered. Finally, you will learn how to integrate Automation hub into the Automation controller so that it consumes the necessary content.

In this chapter, we're going to cover the following topics:

- Overview of Automation hub and its content sources
- Synchronizing certified and community collections
- Publishing custom collections
- Managing execution environments and registries
- Connecting Automation hub to the Automation controller

Technical requirements

This chapter covers collections and EEs and how to manage them using **Configuration as Code** (**CaC**). All the code referenced in this chapter is available at `https://github.com/PacktPublishing/Demystifying-Ansible-Automation-Platform/tree/main/ch09`. You must have Ansible installed to run the code provided.

It is also possible that either the controller or the user's command line won't contain the self-signed certificates from Automation hub. If an error appears stating `x509: certificate signed by unknown authority`, then follow these steps to add the certificate to the host machine:

1. Copy the certificate from Automation hub:

    ```
    $ sudo scp <hub_fqdn>:/etc/pulp/certs/root.crt /etc/pki/
    ca-trust/source/anchors/automationhub-root.crt
    ```

2. Update trusted certificates:

    ```
    $ sudo update-ca-trust
    ```

Overview of Automation hub and its content sources

Automation hub is a repository for the collections and container images that the Automation controller uses. As discussed, in organizations, a credential is used to determine which repository to pull those collections from. Automation hub contains many sources it can pull container images and collections from.

The following diagram shows the sources that are stored in a private Automation hub:

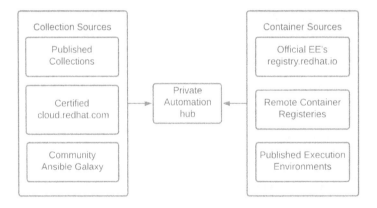

Figure 9.1 – Private Automation hub content sources

Automation hub has three sources for collections, as follows:

*   **Certified**: Certified collections that have been vetted by Red Hat. They are developed, tested, and supported by Red Hat. For some collections, this is in partnership with other companies.

*   **Published**: Collections that have been published to Automation hub. These are collections that anyone who has been granted access to a namespace can publish. These are collections made by you or other team members who share Automation hub.

- **Community**: These are public collections from `https://galaxy.ansible.com`. Anyone can publish to Galaxy and it includes a wide range of collections. Some are modules and roles that have been accepted in the official Ansible distribution but are not supported by Red Hat, while some come from authors respected in the community. It is important to know where the collection or role comes from before using it.

In addition, Automation hub has three categories that EEs fall into:

- **registery.redhat.io EEs**: These are the EEs that ship with Ansible Automation Platform. They currently come in three flavors:

  - **ee-29-rhel8**: Contains the latest version of Ansible 2.9 and all of its dependencies

  - **ee-minimal-rhel8**: Contains the latest supported version of ansible-core and its dependencies

  - **ee-supported-rhel8**: Contains the latest supported version of ansible-core and some of the certified collections come preinstalled

- **Published EEs**: These are EEs that have been published by authorized users of Automation hub. These can be those that you have made, or that other people with access to your Automation hub have published.

- **Remote Registry EEs**: These are EEs that come from remote registries. They could be from Quay.io, Docker Hub, or any other container registry. Because they are coming from a third-party source, they should be scrutinized, but they could also be used to link Automation hub to an internal company registry.

During installation, Automation hub pulls the default images from the container registry; that is, `https://registry.redhat.io`. However, no other content will be preloaded. To load other content, you can use the GUI or the `redhat_cop.ah_configuration` collection, which is designed to interact with Automation hub, just like `ansible.controller` and `redhat_cop.controller_configuration` are built to interact with the Automation controller.

For collections, there are three different sources: the official Red Hat Automation Hub at `cloud.redhat.com`, Ansible Galaxy at `galaxy.ansible.com`, and custom collections uploaded by users.

Synchronizing certified and community collections

The community repository pulls from `galaxy.ansible.com` using a `requirements.yml` file. This allows you to control and curate which public collections can be used. It takes on the following forms and uses name and version information:

```
collection/requirements.yml
    ---
```

```
collections:
  - name: redhat_cop.ah_configuration
  - name: redhat_cop.tower_configuration
...
```

A file source can be an actual file or a URL that leads to a `tar.gz` file.

Documentation about the `requirements.yml` file can be found at `https://docs.ansible.com/ansible/latest/user_guide/collections_using.html#install-multiple-collections-with-a-requirements-file`.

To synchronize collections from either the public Automation hub or Ansible Galaxy, perform the following steps in the GUI:

1. Log in to the Automation hub.
2. Navigate to **Collections | Repository Management | Remote**.
3. For either the community or RH-certified repository, click the three dots on the far right of the page and click **Edit**.

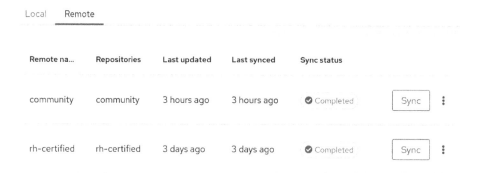

Figure 9.2 – Private Automation hub repository management

4. For each of these, there are options that allow Automation hub to go through a proxy or limit the speed that they're downloaded at. These options are as follows:

   - **Proxy URL | Proxy username | Proxy password**: This is used for proxy settings.
   - **Client Key | Client Certificate | CA Certificate**: This is for authenticating against the proxy.
   - **TLS Validation**: This enforces or ignores TLS validation.

- **Rate Limit**: The requests per second for downloads.

- **Download Concurrency**: The number of collections to download concurrently.

5. For the community repository, the following options are available:

- **URL**: This should be `https://galaxy.ansible.com/api/`, though it can be changed to another Automation hub as an option.

- **Username | Password**: This option isn't needed for `galaxy.ansible.com`, but it can be set for another Automation Hub.

- **Requirements File**: This is a `requirements.yml` file that contains a list of community collections to sync. This file is used for syncing with Ansible Galaxy, as described previously.

6. For the RH-certified repository, there are the following options:

- **URL | SSO URL**: The URL for the repository and the single sign-on URL that's needed to authenticate. These options should be prefilled.

- **Username | Password**: Either the token or username/password must be used to authenticate to the certified repository. This is your username/password for your Red Hat account on `cloud.redhat.com`.

- **Token**: This token value can be retrieved by navigating to `https://cloud.redhat.com/` and then **Logging in | Ansible Automation Platform | Automation Hub | Connect to Hub | Load token**.

7. Click **Save** to save the options. Then, click **Sync** to sync the collections to Automation hub.

These steps can also be completed using the `redhat_cop.ah_configuration` repository. To use the modules in that repository, the following options must be used for each module. An example of this is the method that syncs the community repository from a file, including the necessary authentication variables:

```
//collection/repos.yml
---
- name: Configure community repo from file
  redhat_cop.ah_configuration.ah_repository:
    name: community
    url: https://galaxy.ansible.com/api/
    requirements_file: requirements.yml
    ah_host: https://ah.node
    ah_username: admin
    ah_password: secret123
```

```
    ah_path_prefix: galaxy
    validate_certs: false
...
```

The options for all the modules in the ah_configuration collection use the following options for authentication:

- ah_host: The URL of the host to connect to.

- ah_username: The username to use to authenticate.

- ah_password: The password to use to authenticate.

- ah_token: The token to use in place of a username/password.

- ah_path_prefix: The API path prefix used to connect to Automation hub. For Automation hub, this is galaxy, while for a galaxy_ng installation this is automation-hub.

- validate_certs: This specifies whether or not to validate certificates when connecting to Automation hub.

For the community repository, the requirements.yml file can be supplied or a list of collections to include. You can do so by setting the values for both the RH-certified and community repositories using the same authentication options previously. In the following code, the repositories have been set and synced using the modules for ah_repository and ah_repository_sync. This excerpt can be found in this chapter's GitHub repository. The following example uses a list of collections as opposed to a file:

```
//collection/repos.yml
---
- name: Configure community repo
  redhat_cop.ah_configuration.ah_repository:
    name: community
    url: https://galaxy.ansible.com/api/
    requirements:
      - name: redhat_cop.ah_configuration
      - name: redhat_cop.tower_configuration
        version: 2.0.0
```

The following code will configure the certified repository to gather collections from the Red Hat cloud:

```
- name: Configure rh-certified repo
  redhat_cop.ah_configuration.ah_repository:
```

```
    name: rh-certified
    url: https://cloud.redhat.com/api/automation-hub/
    auth_url: https://sso.redhat.com/auth/realms/redhat-
external/protocol/openid-connect/token
    token: aabbcc
```

To get the token for access to Red Hat Automation hub, follow these steps:

1. Log in to `console.redhat.com`.

2. Navigate to **Ansible Automation Platform** | **Automation Hub** | **Connect to hub** | **Load Token**.

3. Copy the token from here.

Once defined, both the certified and community repositories can be synchronized so that they can grab any updates to collections. This is useful as it puts you into a timed event so that you can periodically update the repositories:

```
- name: Sync community repo
  redhat_cop.ah_configuration.ah_repository_sync:
    name: rh-certified
    wait: true
...
```

So far, we have dealt with collections that come from other sources. The certified and community collections involve other sources that have been written and published. However, the process of publishing collections that you or someone else in your company has written is different. The next section handles how to manage custom collections that haven't been published elsewhere.

Publishing custom collections

For custom collections, a namespace is a prerequisite. Namespaces are the first part of a collection's name – for example, for the `redhat_cop.ah_configuration` collection, `redhat_cop` is the namespace. The namespace is created in the GUI using the following options:

- **Name**: Name of the namespace.

- **Owners**: The group that owns the namespace. It should have the following permission options:

  - `"change_namespace"`

  - `"upload_to_namespace"`

> **Note**
>
> The namespace owner does not have permission to approve a collection. If approval is required, either the namespace owner must be trusted to approve all collections or someone with the approval role must approve updates to collections in the namespace.

- **Company**: Name of the company (optional).

- **Email**: The email used to contact about the collection.

- **Avatar** or **Logo URL**: A URL to a PNG of the image to use for the namespace.

- **Description**: The description of the namespace.

- **Resources**: The documentation of the namespace in Markdown format.

To create a namespace in the GUI, navigate to **Collections | Namespaces | Create**. You will be prompted for the **Name** and **Owners** details. To fill in the other options after creation, navigate to the namespace by going to **View Collections** and click the three dots in the right-hand corner. This will allow you to edit the namespace with the other options.

The format for a list of namespaces in the `ah_configuration` collection is as follows:

```
//collection/namespace.yml
---
ah_namespaces:
  - name: community_test
    company: Community Test
    email: user@example.com
    avatar_url: https://github.com/ansible/awx-logos/blob/
master/awx/ui/client/assets/logo-header.svg
    description: string
    resources: "# Community\nA Namespace test with changes"
    links:
      - name: "Nothing"
        url: "https://github.com/ansible/awx"
    groups:
      - name: ansible_admins
        object_permissions:
          - "change_namespace"
          - "upload_to_namespace"
  ...
```

The role to create a namespace using this format is `redhat_cop.ah_configuration.namespace`.

Once a namespace has been created, a collection can be uploaded by navigating to the collection and clicking the **Upload collection** button.

There are two ways to do this using the `ah_configuration` collection. One involves using the role with a collection list using a Git repository, while the other involves using a list of tarballs of precompiled collections. Here, you can use the following YAML formats:

```
//collection/collection_list.yml
---
ah_collections:
  - collection_name: cisco.iosxr
    git_url: https://github.com/ansible-collections/cisco.iosxr
  - collection_name: awx.awx
    collection_local_path: /var/tmp/collections/repo
ah_collection_list:
  - /var/tmp/collections/awx_awx-15.0.0.tar.gz
...
```

This role uses the following options:

- **Collection name**: The name of the collection
- **Git URL/collection local path**: The location where you can find the collection to compile

You can push the list of collections by invoking the `redhat_cop.ah_configuration.publish` role. This role requires the authentication variables mentioned previously and allows you to automatically approve published collections if needed if you set the `ah_auto_approve` variable to `True`.

Using both the namespace and publish roles, a collection can be published with `collection/publish.yml`. The first step is to make sure the namespace has been created:

```
- name: Create namespace
  include_role:
    name: redhat_cop.ah_configuration.namespace
```

The second step takes the collection list and publishes those collections to Automation hub:

```
- name: Publish Collections
  include_role:
```

```
    name: redhat_cop.ah_configuration.publish
  vars:
    ah_auto_approve: true
...
```

Alternatively, with a compiled tarball, the module can be used to publish a collection. This can be done with the `ah_collection` module:

```
//collection/publish.yml
---
- name: Upload collection to automation hub
  redhat_cop.ah_configuration.ah_collection:
    namespace: awx
    name: awx
    path: /var/tmp/collections/awx_awx-15.0.0.tar.gz
...
```

This is useful when you're publishing the collection from a repository using a CI/CD workflow. Both using the roles and the module are valid. They are designed to be flexible to accommodate either publishing collections from a list of repositories, or folders, or for each repository to publish its collection.

It is best to decide on a strategy for each collection to publish to Automation hub by using the CI/CD pipeline where they are stored. Custom-created collections will have sporadic updates, depending on when the code needs to be updated.

So far, we have covered putting collections in Automation hub. However, Automation hub also acts as a container registry. This allows you to curate and manage EEs. The second half of this chapter will explore how to manage this portion of Automation hub.

Managing execution environments and registries

Managing EEs in Automation hub focuses on two things: grabbing images from a remote registry and managing images that have been pushed to Automation hub. Upon its initial installation, Automation hub does not have any remote registries configured.

How to add a remote EE registry

A remote container registry is easy to interact with. If needed, authentication credentials are provided; this is the base for adding a remote container.

Two popular registries you should add are as follows:

- **Quay.io**: This is a public registry. One of the images you should use here is the EE for AWX (`https://quay.io/repository/ansible/awx-ee`).

- **registery.redhat.io**: This is the official container registry for Red Hat. Automation hub will automatically index any EEs listed there.

For a remote registry, the following fields are used, though only `name` and `url` are required:

- `name`: The name of the registry to remove or modify

- `url`: The URL of the remote registry

- `username`: The username to authenticate to the registry with

- `password`: The password to authenticate to the registry with

- `tls_validation`: Whether to validate TLS when connecting to the remote registry

- `proxy_url`: The proxy URL to use for the connection

- `proxy_username`: The proxy URL to use for the connection

- `proxy_password`: The proxy URL to use for the connection

- `download_concurrency`: Number of concurrent collections to download

- `rate_limit`: Limits the total download rate in requests per second

To set a remote registry in the GUI, follow these steps:

1. Authenticate to Automation hub.
2. Navigate to **Execution Environments** | **Remote Registries** | **Add remote registry**.
3. Fill the form with variables.
4. Click **Save**.
5. For the Red Hat registry, by clicking the three dots at the end, you can view the available index execution environments:

Figure 9.3 – Automation hub remote registry

To set a remote registry using modules, use the following task from the file:

```
//ee/registry_module.yml
---

    - name: Configure Red Hat registry
      redhat_cop.ah_configuration.ah_ee_registry:
        name: rh-registry
        url: registry.redhat.io
```

The module takes the same inputs mentioned previously. The role will also take a list item of these variables:

```
//ee/registery_module.yml
---

    ah_ee_registries:
      - name: quay
        url: https://quay.io/
```

You should use the following role with this variable:

```
    roles:
      - redhat_cop.ah_configuration.ee_registries
```

That covers the basics for adding EEs sourced from an external registry. With a source set, they can be regularly synchronized.

How to synchronize an EE registry

Synchronizing an EE allows you to get updates from the remote registry with Automation hub.

Using the UI, on the same page that an EE registry was added to, click the **Sync from registry** button, as shown in *Figure 9.2*.

To do this with modules, use the following task in Ansible:

```
//ee/registery_module.yml
---

    - name: Sync rh-registry without waiting
      ah_ee_registry_sync:
        name: rh-registry
        wait: false
```

The `redhat_cop.ah_configuration.ee_registries_sync` roles takes the `ah_ee_registries` list input to run the sync module as well.

How to add a remote or local EE to Automation hub

Automation hub acts as its own registry. This means that containers can be pushed to it using commands, or by the hub pulling containers from remote registries.

To push a local container to Automation hub, follow these steps. You can review the commands mentioned here by rereading *Chapter 8, Creating Execution Environments*:

1. Retag the image with the remote registry in its name. You can also rename the image here:

    ```
    $ podman tag localhost/base_ee ah.node/base_ee
    ```

2. Push the new image to Automation hub:

    ```
    $ podman push --tls-verify=false ah.node/base_ee
    ```

Alternatively, you can use the `redhat_cop.ee_utilities` roles to push EEs. A third option is to use the `containers.podman.podman_image` module.

For a remote execution environment, the following fields are used. Only `name`, `upstream_name`, and `registry` are required:

*   `name`: The name of the EE to use. This can be different from the remote name.
*   `upstream_name`: The name of the EE in the remote registry.
*   `registry`: The remote registry to pull from, including the repository; for example, `ansible/awx-ee`.
*   `tags_to_include`: Tags to include from the external registry.
*   `tags_to_exclude`: Tags in the external registry to exclude.
*   `description`: Description of the EE.
*   `groups`: Groups that have access to containers in the namespace. In this example, an EE called `cheddar/awx-ee` would grant access to all the containers in the `cheddar` namespace.

To add an EE from a remote registry to the GUI, follow these steps:

1. Authenticate to Automation hub.
2. Navigate to **Execution Environments | Execution Environments | Add execution environment**.
3. Fill in the form with variables.
4. Click **Save**.

At the time of writing, there isn't a module or role for adding an EE to Automation Hub with a role or module. However, the `redhat_cop.ah_configuration` collection may have created one after this book is published.

Now that we've added remote registries and EEs, let's learn how to maintain them.

Best practices for maintaining execution environments

It is a best practice to regularly update EEs, similar to synchronizing the registry.

With `registry.redhat.io` added, synchronized, and indexed, the following execution environments should auto-populate:

Container repository name ↓	Descri...	Created ↕	Last modifi...	Contain...	
ansible-automation-platform-21/ee-29-rhel8	Ansibl...	6 days ago	6 days ago	Remote	⋮
ansible-automation-platform-21/ee-minimal-rhel8	Ansibl...	5 days ago	5 days ago	Remote	⋮
ansible-automation-platform-21/ee-supported-rhel8	Ansibl...	5 days ago	5 days ago	Remote	⋮

Figure 9.4 – Automation hub execution environments

These are the official execution environments that are regularly updated with bug fixes and **Common Vulnerabilities and Exposures** (**CVEs**) updates.

Using a CI/CD to build execution environments, as discussed in *Chapter 8, Creating Execution Environments*, allows for regular updates of EEs when new versions of the base image are released. There is no notification or set schedule for the updates, however the current cadence appears to be every 2-3 weeks.

It is also recommended to either match the EE tag with the one it was built with, or to have that noted somewhere that it can be accessed. This allows for easier troubleshooting if issues arise.

All of these are puzzle pieces. For example, there is a source for base images. It is better to use the EEs from the Red Hat registry as they can be updated versus the static ones that the automation controller has upon being installed. From there, the customized EEs provide more flexibility for different tasks in the controller. However, Automation hub is merely a repository for information that can be used in the controller. The next step is to transfer that information to the Automation controller.

Connecting Automation hub to the Automation controller

The Automation controller provides mechanisms for consuming content from Automation hub. These include both collections and execution environments.

In *Chapter 5, Configuring the Basics after Installation*, we covered credentials and organizations – specifically, Galaxy credentials. When the Automation controller and Automation hub are installed together, this creates a default **Automation Hub container registry** credential. If this is not present, a **container registry** and **Ansible Galaxy** credential are used to authenticate the controller to Automation hub.

Using credentials to pull collections

The Ansible Galaxy credentials are used in an organization on an automation controller, the order of which determines which collections are pulled. It is recommended that you use the Published, Certified, Community, Galaxy order. This ensures that any self-published collections are pulled first, that a certified collection is prioritized over community ones, and that previously synced community collections are pulled last.

When a project is synchronized on the Automation controller, it uses a `collections/requirements.yml` file to list collections to install alongside the project. This is the same file that we mentioned when building EEs. These take precedence over those installed in the EE. The Automation controller will use the Galaxy credentials in the organization to find and download any collection that's required.

Adding execution environments to the Automation controller with the GUI

To use execution environments, they must be added to the Automation controller. These can then be used in organizations, projects, inventories, and job templates. Automation hub has a shortcut for adding EE to the controller in the GUI.

To add an EE from Automation hub to the Automation controller, follow these steps:

1. Authenticate to Automation hub.

2. Navigate to **Execution Environments | Execution Environments**. Then, click on the three dots to the right of the EE you want to add. Next, click **Use in Controller** and click on the controller's URL; for example, `https://controller.node`. After that, authenticate to the Automation controller; a form will appear so that you can add the EE:

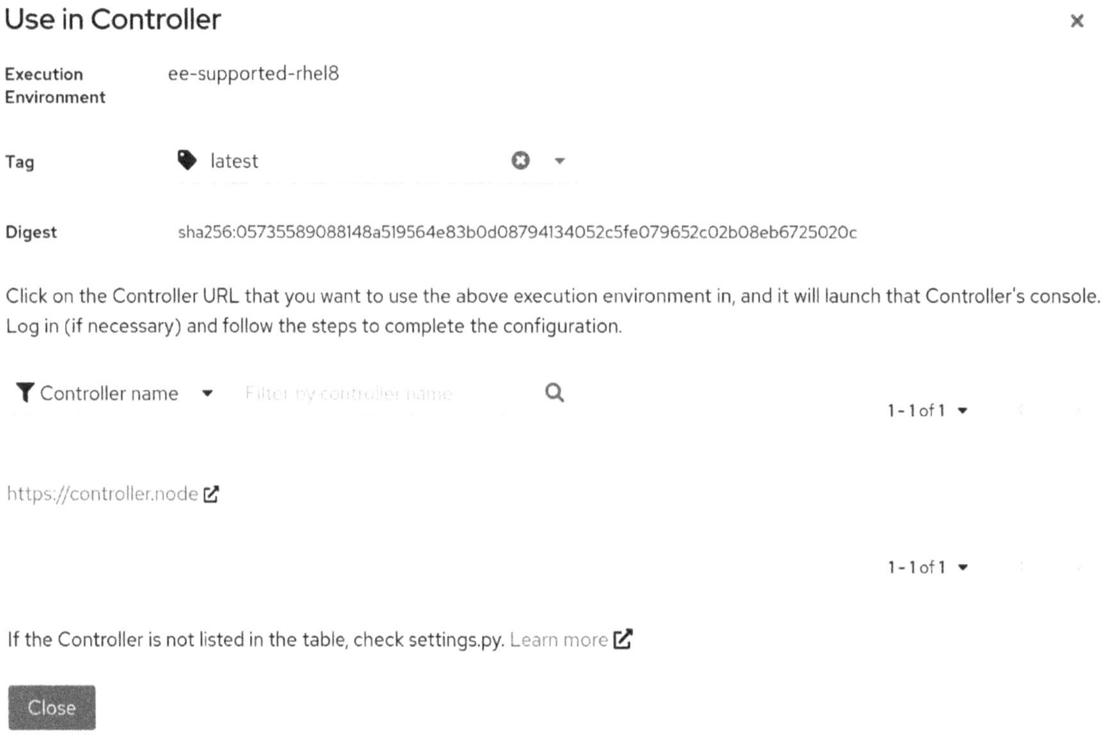

Use in Controller   ✕

Execution   ee-supported-rhel8
Environment

Tag   🏷 latest   ✕ ▾

Digest   sha256:05735589088148a519564e83b0d08794134052c5fe079652c02b08eb6725020c

Click on the Controller URL that you want to use the above execution environment in, and it will launch that Controller's console. Log in (if necessary) and follow the steps to complete the configuration.

▼ Controller name ▾   Filter by controller name   🔍   1 - 1 of 1 ▾

https://controller.node 🔗

                                                                                    1 - 1 of 1 ▾

If the Controller is not listed in the table, check settings.py. Learn more 🔗

Close

Figure 9.5 – Automation hub – Use in Controller

This can also be done in the controller UI, as follows:

1. Authenticate to the Automation controller.
2. Navigate to **Execution Environments** and click **Add**.

The following fields are used to add EEs:

- **Name**: The name to use for the EEs.
- **Image**: The URL of the container image to use, including the tag. For example, you could use `ah.node/ansible-automation-platform-21/ee-minimal-rhel8:latest`.
- **Description**: Description of the EE.
- **Organization**: The organization the EE is limited to. Leave this blank to make it globally available.
- **Credential**: The credential to use to authenticate to the container registry.
- **Pull**: What the pull behavior should be. Here, you have the following options:
  - **Always**: Always pull the EE image.

- **Missing**: Only pull the EE when it's missing.

- **Never**: Never pull the EE image.

With the EE in the Automation controller, it can be used for tasks. However, this is not the only way to add an EE. The other option is to use modules.

Adding execution environments to the Automation controller with modules

Using a module to add an EE is useful for a playbook where you create an EE, publish it to a registry such as Automation hub, and then add it to the controller.

The following is an excerpt from the `controller/controller_ee_modules.yml` file:

```
- name: Create EE
  ansible.controller.execution_environment:
    name: "My EE"
    image: quay.io/ansible/awx-ee
```

The module uses the same fields as the GUI does, only with the module option in lowercase. While the module is useful for adding a single execution environment, maintaining it is best done with the configuration role.

Adding execution environments to the Automation controller with roles

Using a role to add and manage EEs on the controller is useful for maintaining CaC. The following is an excerpt from the `controller/configs/execution_enviroments.yml` configuration file:

```
controller_execution_environments:
  - name: "My EE"
    image: quay.io/ansible/awx-ee
    pull: always
```

The role list uses the same fields as the GUI does, only with the module option in lowercase.

The `redhat_cop.controller_configuration.execution_environments` role takes the configuration file as input and adds the EEs to the Automation controller.

With the options to add EEs to the Automation controller with the GUI, modules, and roles, EEs can be used with jobs, projects, and inventories.

Summary

In this chapter, you learned how to manage Automation hub, as well as about the collections and EEs that are used in job templates on the Automation controller. Keeping execution environments and collections up to date is important so that playbooks can be run and kept up to date.

The previous chapters covered projects, inventories, credentials, and **role-based access control (RBAC)**. All of those pieces are related to job templates, which run playbooks inside the Automation controller. Workflows are a series of job templates. Projects, workflows, and job templates will be covered in the next chapter.

Creating Job Templates and Workflows

Before now, this book has covered many different parts of the Automation controller, inventories, credentials, execution environments, and collections. These are all pieces that go into a job template. Job templates are what the Automation controller uses to run Ansible playbooks. Workflows are job templates linked together in a logical flow. This chapter will explore the details of how to create projects that the job templates use, these job templates themselves, how to create workflows, and the basics of using them.

This chapter will also cover how to incorporate playbooks and code into the Automation controller using projects, create job templates, use surveys in job templates and workflows, create workflows, and use instance groups.

In this chapter, we're going to cover the following main topics:

- Creating projects
- Creating job templates
- Surveys for job templates and workflows, and how to use them
- Creating workflows
- Using job slicing to slice a job template into multiple jobs

Technical requirements

This chapter will go over collections, execution environments, and how to manage them using **Configuration as Code** (**CaC**). All the code referenced in this chapter is available at https://github.com/PacktPublishing/Demystifying-Ansible-Automation-Platform/tree/main/ch10. It is assumed that you have Ansible installed in order to run the code provided. It is also assumed that you have installed the Automation controller and Ansible Automation hub.

Creating projects

Projects are what we use to load playbooks into the Automation controller. This can be a file loaded on the controller filesystem or a Git repository. The controller will search for any valid playbook files in the project so that they can then be referenced by job templates. Project files should follow the standard playbook layout.

An important thing to keep in mind with projects and playbooks is not to try and create an all-encompassing project that contains every playbook or task file in a single project. There should be forethought and planning for how it is best to divide the playbooks. If everything goes into one project repository, it will eventually become too big to manage even as a team. This can lead to bloat, unused code, and cause issues when trying to remove one piece of the code, which leads to other things breaking.

This is also why collections and roles are important. Collections can contain modules and groups of roles in a cohesive unit. Separating the code from the project allows for the portability of code between projects. When starting out, it's acceptable to use a single project, but keep these ideas in mind as it grows, so that its size does not get out of hand.

The structure and files of a playbook directory

An example of the playbook format exists in this chapter's repository files. The `projects/playbook_dir` folder and the following file structure will reference `projects/playbook_dir_struct.yml`.

This file and folder structure looks as follows:

```
├── collections
│   └── requirements.yml
├── roles
│   └── requirements.yml
├── tasks
│   └── tasks.yml
├── playbook.yml
└── ansible.cfg
```

Figure 10.1 – The playbook directory structure for projects

When a project syncs, it looks for the following files:

```
collections/requirements.yml
roles/requirements.yml
```

The `requirements` file for both collections and roles are evaluated separately for items to install. For example, roles in the `collections` file will not be installed. The format of these was discussed in *Chapter 9, Automation Hub Management*. More information about the format can be found here: `https://galaxy.ansible.com/docs/using/installing.html`. This folder and file needs to go in the base directory of the project.

In addition, the `collections` folder and the `roles` folder can also go in here, but it is recommended to store and pull them from Galaxy. However, when developing new roles or collections, it is sometimes easier to just keep them alongside the playbook. This is done by housing the collection in the `collections/ansible_collections` playbook folder. More information about using this method can be found here: `https://docs.ansible.com/ansible/latest/user_guide/collections_using.html`.

The following are module folders where modules, plugins, and other content specific to the playbook are stored. If you have custom Python code that is not in a module, this is where it should go. These are stored in parallel to the playbook directory, meaning that a playbook that starts in a nested folder will not see these, unless the `ansible.cfg` file tells it to look there:

```
library/
module_utils/
filter_plugins/
```

`playbook` and `tasks` files and folders are going to make up most of the project files:

```
tasks/
playbook.yml
```

There can be multiple task folders and playbook files side by side. It can make sense to store them together, but be conscious of bloat and creating monolithic projects that have dozens of playbooks and task files, as this can make it harder to keep track of code that falls into disuse, and harder to keep the project and code updated. Playbook and task files can be nested into their own folders and subdirectories, which can make organization easier.

One of the most overlooked files is the `ansible.cfg` file. This stores any Ansible configuration changes that are used. This is an optional file. However, when using things such as inventory settings, a Galaxy without valid certificates, or other configuration changes, this is the place to store any configuration settings. The Automation controller will automatically pick this up from the base directory of the project.

A popular configuration setting to use in the `config` file is to turn off certificate validation for Galaxy. This is useful when using self-signed certifications, which can cause issues when syncing roles and collections from Automation hub with certificates that the controller does not recognize. Additional settings can be found here: `https://docs.ansible.com/ansible/latest/reference_appendices/config.html`.

The code to add in the `ansible.cfg` file to turn off certificate validation is as follows:

```
[galaxy]
ignore_certs = True
```

Technically, every one of these files is optional. However, it would not make sense to have a project without a playbook, or something else used, such as an inventory source plugin. It can make sense to include inventory definitions with playbooks, but the best practice is to separate them from the playbooks for easier management.

This section covered how a project directory should be structured and how different files are used. The next step is to create a project object on the Automation controller.

Project options on the Automation controller and creating projects in the GUI

The different options will be broken up into unique groups. GUI fields will be represented with **NameBolded** fields. If the module and role fields are anything other than lowercase, they will be represented with (`namefield`).

Projects can be created in the GUI using the following steps:

1. Navigate to **Projects** on the left-hand sidebar.
2. Click **Add** and fill in the following fields:

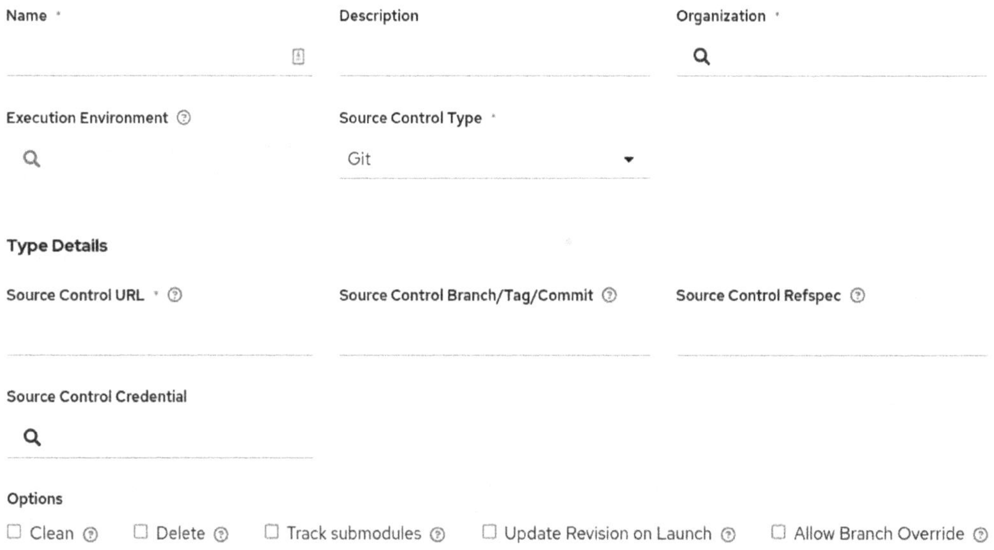

Figure 10.2 – A project creation form

> **Note**
>
> If the module or role option is not the same as the field, it will be noted in a code-style format.

We have filled in the following data:

- **Name**: A unique inventory name within the organization it belongs to.

- **Description**: The description of the project.

- **Organization**: The organization that the project belongs to. Job templates will inherit their project's organization.

- **Execution Environment**: We will set it to `default_enviroment`, which sets the execution environment for jobs of this project to default.

- **Source Control Credential Type**: We will set this to `scm_type`, indicating what type of source control this project uses. These will be covered in more detail as follows.

  - **Manual**: The Automation controller will, by default, look in `/var/lib/awx/projects` for project folders. All controller nodes should have their projects copied there to use this option.

  - **Git**: This clones a Git branch to use as a project directory.

  - **Subversion** (`svn`): clones a subversion revision as a project directory.

  - **Red Hat Insights** (`insights`): is used by playbooks to connect to and use Red Hat Insights.

  - **Remote Archive** (`archive`): is used to access a remote archive such as a GitHub release or Artifactory archive. For example, see the following: `https://github.com/sean-m-sullivan/test-playbooks/archive/refs/tags/v1.0.0.tar.gz`.

3. Choose the additional settings outlined in instructions for each project type detailed next.

4. Click **Save**.

These are the basics of how to create a project using the GUI. However, there are more options to set.

The **Manual** projects have the **Path** option or the `local_path` option to set which folder the playbook directory will be in.

> **Note**
>
> Manual projects should only be used as a method of last resort. They are difficult to maintain as they do not have auto-update or version control. A Git server is recommended instead of this option. GitLab is a good open source option for this purpose.

For every other type, **Git**, **Subversion**, Remote Archive, and Red Hat Insights, there are some shared options:

- **Source Control Credential** (`credential`): The credential to use to authenticate a project source.

- **Clean** (`scm_clean`): Removes any local file changes before an update.

- **Delete** (`scm_delete_on_update`): Removes all local files before an update.

- **Track Submodules** (`scm_track_submodules`): Tracks submodules or subprojects if set in the remote repository.

- **Update Revision on Launch** (`scm_update_on_launch`): Update the revision whenever a job is launched for the project. This can lead to a significant wait between jobs and even cause bottlenecks. It is better to either set a schedule to update or have a Webhook or **continuous integration/continuous delivery (CI/CD)** to force a project update when changes are made.

- **Cache Timeout** (`scm_update_cache_timeout`): The time before the project update is considered old and needs to be updated if **Update Revision on Launch** is set.

For Git, Subversion, and Remote types, but not Insights, refer to the following:

- **Source Control URL** (`scm_url`): The URL for the source control project, such as the following:

  - `https://github.com/sean-m-sullivan/inventory_example.git`

  - Example: `git@github.com:sean-m-sullivan/inventory_example.git`

  - Example: `http://svn.example.com:9834/repos`

- **Allow Branch Override** (`allow_override`): Allows a job template to override the default branch of the project on prompting.

- For the Git and Subversion types: **Source Control Branch/Tag/Commit/Revision** (`scm_branch`): Determines which branch to use for the source control. This can be a variety of options depending on the source, whether branches, tags, or commit hashes.

- For the Git type: **Source Control Refspec** (`scm_refspec`): This allows for more complex references beyond just branch.

These fields also apply to the roles and module use cases. Changing things or creating a project in the GUI is great for testing. A good example is turning on **Update Revision on Launch** when doing iterative testing inside the Automation controller. You can make a change to the playbook code, commit it to a branch, change the branch used in the GUI, and turn on update on launch. Doing this once allows for a cycle of making a change, launching a job, and observing the result, rather than manually updating the project every time.

In addition to creating and changing settings for projects in the GUI, the modules and roles can also be useful for making changes to a project.

Creating projects on the Automation controller using modules

In addition to the preceding options, the modules and roles have the following options to use as well:

- update_project: This determines whether or not to update the project when creating or updating it with the module. This is very useful for creating new projects, as job templates will fail without an initial update or sync, as the Automation controller cannot find the playbook.

- wait: A Boolean that determines whether or not to wait for the project to update before moving on to the next task.

- timeout: If wait is set, this refers to the time in seconds for the update to time out.

- interval: If wait is set, this refers to the interval at which to check whether the update has been completed yet.

- state: This can be set to present or absent. The default is present, but this can be used to remove projects from the controller.

The module can be invoked from the following task, which is an excerpt from the projects/ set_project_using_module.yml file. The full file can be found in this book's repository chapter folder:

```
- name: Create the Test Project
  ansible.controller.project:
    name: Test Project
    scm_type: git
    scm_url: https://github.com/ansible/test-playbooks.git
    scm_branch: master
    scm_clean: true
    description: Test project
    organization: Default
    wait: true
    update_project: true
```

Using the modules for a task to update or create a project can be useful for a quick update. However, maintaining configuration state for the Automation controller is best done with the configuration roles.

Creating projects on the Automation controller using roles

The steps for creating and maintaining projects with roles are as follows. The projects role takes a list of projects with options and applies them to the Automation controller:

```
//projects/set_project_with_roles.yml
    controller_projects:
      - name: Test Project
        scm_type: git
        scm_url: https://github.com/ansible/test-playbooks.git
        scm_branch: master
        scm_clean: true
        description: Test Project 1
        organization: Default
        wait: true
        update_project: true
```

This is the full name of the group role to apply to the projects roles:

```
      - redhat_cop.controller_configuration.projects
```

With projects created and set, all the pieces for creating job templates are in place. The following section will go over the basics of job templates and their creation.

Creating job templates

This section will go over job template creation. Job templates are the heart of the Automation controller. Everything is built around them, which is one of the reasons they have the most options; at last count, they had almost 40.

Job templates control how playbooks run. They control which hosts are used, variables are included, and the how a job behaves when run. This can seem daunting. However, most of the options are *optional*; they are knobs to turn and use as needed.

At the same time, a job template is meant to be a template to run the same thing many times. While things such as surveys and variable inputs can change how a particular playbook runs, there is no reason why a playbook can't be used in multiple job templates.

In Ansible, there are always multiple ways to achieve the same goal. This applies to job templates as well. Throughout this chapter and the next chapter, with workflows and job templates, there are various strategies for using job templates effectively. To utilize these strategies, understanding the options is essential.

Job template options for the GUI, modules, and roles

The different options will be broken up into unique groups. GUI fields will be represented with **NameBolded** fields. If the module and role fields are anything other than lowercase, they will be represented with (`namefield`):

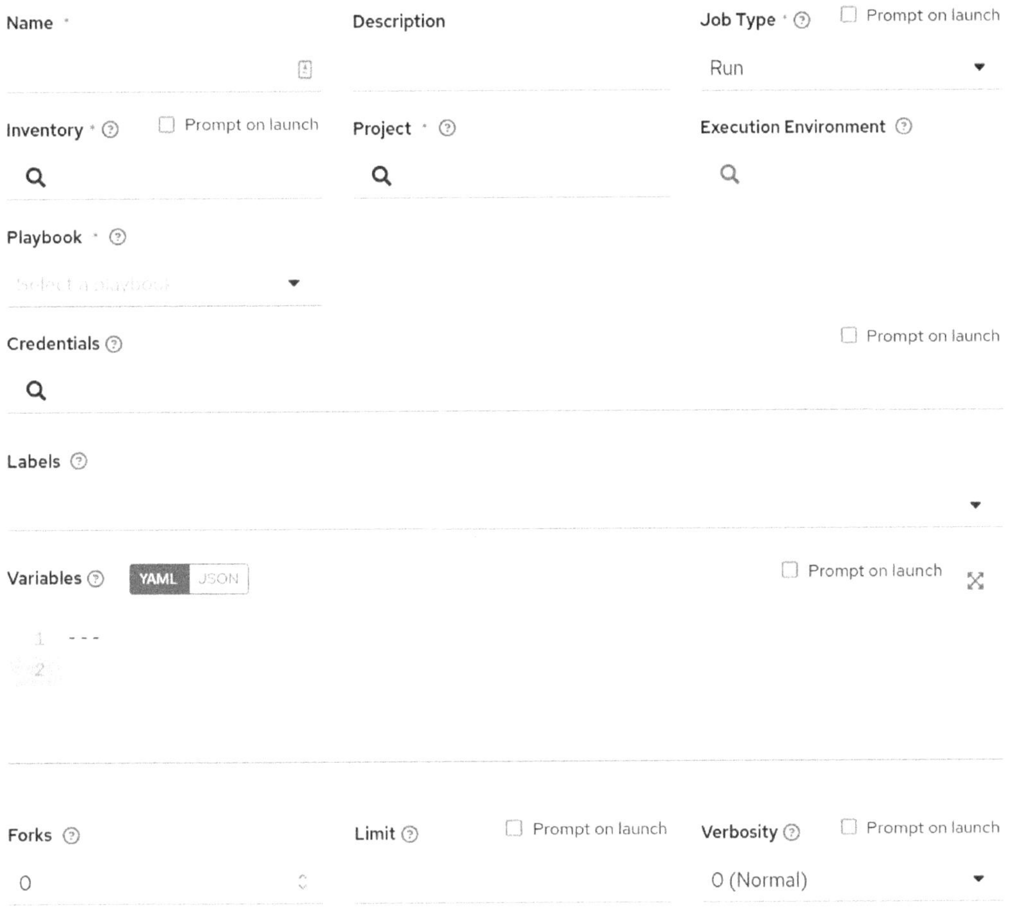

Figure 10.3 – Partial job template creation form

The fields required to make a playbook run are **Name**, **Inventory**, **Project**, and **Playbook**. The rest are all optional variables.

The basic options that are most used for creating a job are the following:

- **Name**: The name of the job template.
- **Description**: The description of the job template.

- **Job Type** (**Run** or **Check**): **Run** is the default behavior and runs the job making changes. **Check** makes no changes and runs tasks in **Check** mode.

- **Inventory**: The inventory to use for the job template.

- **Project**: The project to use.

- **Execution Environment**: The execution environment to use.

- **Playbook**: The playbook path to use in the project.

- **Credentials**: A list of credentials to use in the job.

- **Variables** (`extra_vars`): The key-value dictionary pairs of variables to use in the job.

Some of the more advanced options you can use in a job template are as follows:

- **Forks**: The number of processes to use when executing the playbook. The default with Ansible is five. This can either hinder or speed up a running job.

- **Limit**: A set of host patterns to limit the hosts from the inventory that will be included in the job.

- **Verbosity**: Ranging from one to four, this is how verbose the debug output will be for the job. It is useful when diagnosing problems with the job.

- **job_slice_count**: The number of slices to use for the job. This will effectively turn the job template into a workflow. This will be further covered later in the *Using job slicing to slice a job template into multiple jobs* section.

- **Timeout**: The duration of time the job runs for before it is canceled. Defaulting it to zero disables it.

- **diff_mode**: This shows the changes made by Ansible tasks, effectively running Ansible with the `--diff` option.

- **Source Control Branch** (`scm_branch`): This defines which branch of the project the job template should run on. The `allow branch override` project option must be set.

Some of the lesser-used options for running a playbook are the following:

- **Enable Fact Storage** (`use_fact_cache`): This will store gathered system facts about the host in the inventory. These can then be used on later jobs.

- **Instance groups** (`instance_groups`): This defines which instance or node the job template should run on. This is covered later in *Chapter 14, Automating at Scale with Automation Mesh*.

- **Job tags** (`job_tags`): If tasks in Ansible are tagged, this defines which tags to use in the job.

- **Skip tags** (`skip_tags`) If tasks in Ansible are tagged, this defines which tags to skip in the job.

- **Privilege Escalation** (`become_enabled`): This enables privilege escalation – this is the same as using `--become` in an Ansible command-line.

- **Labels**: The labels to apply to a job template – this can be used to group and filter job templates either through the GUI or through queries. In the GUI, the labels are created automatically. If using modules, roles, or API methods, the label must be created first.

- **Concurrent Jobs** (`allow_simultaneous`): This allows several instances of the job template to run at once.

- **Provisioning Callbacks** (`host_config_key`): This is a key for hosts to use to call this job template to run against themselves for provisioning. More details can be found here: `https://docs.ansible.com/automation-controller/latest/html/userguide/job_templates.html#provisioning-callbacks`.

Another set of options are prompts. These will prompt the user for inputs to override when used to launch a job or used in a workflow. They are usually checkboxes labeled **Prompt on launch** in the GUI, as shown in the following figure:

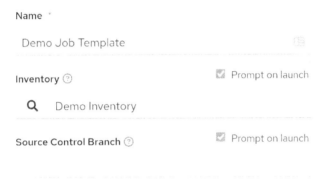

Figure 10.4 – Prompt on launch checkboxes

The prompts are the following:

- `ask_diff_mode_on_launch`: This will prompt the user to clarify which differential mode to use.

- `ask_variables_on_launch`: This will prompt the user to input extra variables to use.

- `ask_limit_on_launch`: This will prompt the user to enter a limit.

- `ask_tags_on_launch`: This will prompt the user to enter or temporarily remove the tags to use in the job.

- `ask_skip_tags_on_launch`: This will prompt the user to enter or temporarily remove the tags to skip in the job.

- `ask_job_type_on_launch`: This will prompt the user to clarify which job type to use on launch.

- `ask_verbosity_on_launch`: This will prompt the user to clarify which verbosity to use on launch.

- `ask_inventory_on_launch`: This will prompt the user to clarify what inventory to use on launch.

- `ask_credential_on_launch`: This will prompt the user to add additional credentials on launch.

The following options are only used in the modules, the role, or the API:

- `force_handlers`: If the task fails, it forces the handlers to run.

- `start_at_task`: The name of the task to start on.

The following options are only used in the modules or the role:

- `organization`: The organization to use to look up the job template – this does not change what organization the template is in. This is handy if multiple organizations use the same job template name, as, without it, multiple templates will be returned, causing a task failure.

- `state` (`present` or `absent`): This describes the state of the job template. This is useful for removing job templates.

In addition, there are two options for surveys, which will be discussed further in the *Surveys for job templates and workflows, and how to use them* section:

- **survey_enabled**: For turning the survey prompt on and off.

- **Survey Question** (`survey_spec`): These are survey questions to use.

With the options to use laid out, we will now go on to learn about using these options to create job templates through the GUI, modules, and roles.

Creating job templates using the GUI

Job templates can be created in the GUI using the following steps:

1. Navigate to **Templates** on the left-hand sidebar.
2. Click on **Add**, select **Add job template**, and fill in the **Name**, **Inventory**, **Project**, and **Playbook** fields as required.
3. Click **Save**.

Making changes in the GUI can be useful for making last-minute changes, testing changes, or making tweaks. For example, the verbosity level of the job can be mutable. When troubleshooting, it can be very useful to increase it to three temporarily to see the additional output, any errors, and data on what is happening in the job.

Creating job templates using modules

The module can be invoked from the following task; it is an excerpt from the `//job_templates/set_job_template_using_module.yml` file. The full file can be found in this book's repository chapter folder:

```yaml
- name: Create job Template
  Ansible.controller.job_template:
    name: test-template-1
    description: created by Ansible Playbook
    project: Test Project
    inventory: RHVM-01
    playbook: chatty_payload.yml
    credentials:
      - admin@internal-RHVM-01
    verbosity: 2
    extra_vars:
      target_hosts: infra-ansible-01.example.com
    job_type: run
    state: present
```

Using the modules to update a task or create a job template can be useful for a quick update, or even for creating a dynamic job template to launch with unique changing variables. However, maintaining the state of the Automation controller is best done with the configuration roles.

Creating job templates using roles

The steps for creating and maintaining job templates with roles are as follows. The job template's role takes a list of job templates with options and applies them to the Automation controller:

```yaml
//job_templates/set_job_template_with_roles.yml
  controller_templates:
    - name: test-template-1
      description: created by Ansible Playbook
      job_type: run
      inventory: RHVM-01
      credentials:
        - admin@internal-RHVM-01
      project: Test Project
```

```
playbook: chatty_payload.yml
verbosity: 2
extra_vars:
  target_hosts: infra-ansible-tower-01.example.com
```

This is just an excerpt from the file. The other pieces that are required to make sure that all the pieces are created are available for reference in the full file. This will fail if the inventory, credential, or project is not created first.

The recommended list of roles to use to make sure all the pieces are in place is as follows:

- redhat_cop.controller_configuration.credentials
- redhat_cop.controller_configuration.inventories
- redhat_cop.controller_configuration.projects
- redhat_cop.controller_configuration.job_templates

The file example also includes the configuration pieces for the preceding roles to be invoked alongside the job template.

That covers the creation of job templates, their options, and multiple methods of creation. The following section will cover surveys. Surveys are used in job templates and workflows to gather user input. How a playbook runs is determined by variables and survey variables. The following section will go into detail about how to create surveys and use them in both job templates and workflows.

Surveys for job templates and workflows, and how to use them

Surveys are a way to capture user input for job templates and workflows. Each question has a set variable that is populated for use by playbooks. This is also useful when users are using the API, roles, or modules. Surveys allow users to input fields as extra variables into the job. When using only surveys this limits users to only the extra variables defined in the survey. Those fields must match the data type set in the survey, which adds a level of data validation. With that capability in mind, the next section reviews how to define surveys.

Survey definitions

The top-level options for surveys in modules and roles for job templates and workflows are the following (survey_spec):

- name: The name of the survey – this does not show up in the GUI.
- description: The description of the survey – this does not show up in the GUI.
- spec: A list of survey questions.

All three of these are required when creating a survey using roles or modules. In addition, any survey questions not in the `spec` list when posted will be removed.

Each `spec` list item is a survey question. Each of these questions can have a different type of answer. There are seven different answer types. They consist of the following:

- `text`: A string, the length of which can be set
- `textarea`: A multi-line string that can also be limited by the number of characters
- `password`: A string – however, in the log and job output, the string is encrypted
- `multiplechoice`: Multiple-choice options that are given to the user – only one can be selected
- `multiselect`: Multiple-choice options that are given to the user – the user can select multiple choices
- `integer`: An integer number input
- `float`: A float number input

This defines what the variable type will be for each question's answer for the playbook to use in the job.

For example, this is the form to create a multiple-choice survey question:

Question *	Description	Answer variable name * ?
multiple_single		multiple_single

Answer type * ?	☑ Required
Multiple Choice (single select) ▼	

Multiple Choice Options * ?

choice1	⊞
choice2	
choice3	
choice4	

Figure 10.5 – A multiple-choice survey form

Each survey `spec` list item contains the following fields:

- `question_name`: The question that will be displayed to the user.
- `question_description`: Description to display as a question mark for the field. Each field has a **?** to click; the text of this field will show there.

- **required**: This defines whether a question in the survey has to be answered or not.

- **type**: This defines the type of variable to use for the question.

- **variable**: This defines the variable name to use that will be accessible in the playbook.

- **min**: This sets the minimum length of the answer.

- **max**: This sets the maximum length of the answer.

- **default**: This sets the default answer to the question if there is one.

These options give you control over the form and how the inputs are used.

For the `multiplechoice` option and the `multiselect` option, there is an additional field called `choices`. This is a list of the choices for the user to select from. However, the `multiselect` option is paired with another field, `formattedChoices`. This allows multiple defaults to be selected, and a list order to be created through the `id` field:

```
formattedChoices:
  - choice: asdf
    isDefault: true
    id: 0
```

A reference document has been made for survey options in this chapter's repository in the `survey_spec.yml` file.

Using surveys

By their nature, surveys are static. They have no dynamic inputs. However, using job template roles, workflow roles, and modules, a survey can be updated. Using variables populated from an API, or another source, a survey's options for users can be updated, making the survey values semi-dynamic. There is a multitude of ways to achieve this, especially because the source of data varies for everyone.

Surveys allow for dynamic, controlled input for users when launching job templates and workflows. When creating a workflow node from a job template, it prompts a survey to set the answers as static.

Surveys allow user input for job templates and workflows. The following section will cover workflow creation. It is important to cover surveys beforehand since workflow nodes can take survey variables as inputs, as they are part of job templates.

Creating workflows

Workflows are job templates that allow a flow of jobs in the Automation controller. They are represented in a visualizer or workflow node list. Workflows can consist of job templates, project syncs, inventory source sync, and other workflows. Each of these items is linked by basic logic gates to determine which item in the flow is executed next.

Returning to job templates and projects, workflows combine job templates, and they do not care which project a job template comes from. A design decision can be made to separate projects with playbooks that are used in multiple workflows. This section will go into the details of creating workflows and the basics of using logic between the different nodes. Putting these items together along with strategies for playbooks, job templates, and nodes will be covered in *Chapter 11, Creating Advanced Workflows and Jobs*.

Workflow basics

Workflows are made up of three basic parts: the workflow template, the survey, and the workflow node list. The survey part is for user input, which we covered in the previous section, *Surveys for job templates and workflows, and how to use them*. The workflow template covers the basics of managing the workflow object in the Automation controller, along with the details of the workflow job. The workflow node list dictates the nodes in a workflow and how logic flows between them.

Workflow job template options for the GUI, modules, and roles

The different options will be broken up into unique groups. GUI fields will be represented with **Name**. If the module and role fields are anything other than lowercase, they will be represented with (name),

The only field required to create a workflow is the **Name** field:

Name *	Description	Organization
Complicated workflow schema		
Inventory ⓘ ☐ Prompt on launch	Limit ⓘ ☐ Prompt on launch	Source control branch ⓘ ☐ Prompt on launch

Labels ⓘ

Variables ⓘ **YAML** JSON   ☑ Prompt on launch

    1 - - -

Options

☐ Enable Webhook ⓘ   ☐ Enable Concurrent Jobs ⓘ

Figure 10.6 – A workflow creation form

The basic options that are most used for creating a job are the following:

- **Name**: The name of the workflow.

- **Description**: The description of the job template.

- **Variables** (`extra_vars`): Key-value dictionary pairs of variables to use in each of the jobs in the workflow.

- **Organization**: The organization where the workflow exists.

- **Concurrent Jobs** (`allow_simultaneous`): Allows multiple instances of the job template to run at once.

- **Inventory**: An inventory to use in jobs with job templates that use **Prompt on launch** options.

- **Limit**: A set of host patterns to limit the hosts from the inventory that will be included in the job.

- **Source control branch** (`scm_branch`): This defines which branch of the project a job template should run on. The `allow branch override` project option must be set. This is only for when a job template has the `ask scm` branch set on launch as well.

- **survey_enabled**: This can be used to turn the survey prompt on and off.

- **Survey Question** (`survey_spec`): These are survey questions to use.

- **Labels**: A label to apply to the workflow job template – this can be used to group and filter job templates, either using the GUI or queries. In the GUI, labels are created automatically. If using modules, roles, or API methods, labels must be created first.

- **Visualizer** (`workflow_nodes`): The list of workflow nodes and their logic. This will be covered in the following section.

Similar to job templates, the workflow job template has several options that are labeled **Prompt on launch** in the GUI. Whether these options are applied to an underlying job template depends on whether the job template has been set to allow a prompt on launch. The prompts consist of the following:

- `ask_scm_branch_on_launch`: This will prompt the user to input the branch to use in the **Source control branch** field. The same limitations apply in that this will only be used if the job template also has this prompt set.

- `ask_variables_on_launch`: This will prompt the user to input extra variables to use.

- `ask_inventory_on_launch`: This will prompt the user to clarify which inventory to use on launch. The same limitations apply in that this will only be used if the job template also has this prompt set.

That covers the options for workflow creation. A workflow can be created with all these options and have no jobs to use when launched. Now that we have defined the options, the next step is to create a workflow job template.

Creating a workflow job template using the GUI

Workflow job templates can be created in the GUI using the following steps:

1. Navigate to **Templates** on the left-hand sidebar.

2. Click on **Add**, select **Add workflow template**, and fill in the fields from the previous section (**Name** is required).

3. Click **Save**.

After this, a workflow job template has been created. However, this leaves the workflow node list to cover. The following section will explore the creation of the workflow nodes and logic.

Workflow node list options and creation in the GUI

A workflow node list describes the logical flow of jobs in the Automation controller using a graph-like structure with nodes. A workflow node list is best described with a graphic as follows:

Figure 10.7 – Visual representation of a workflow

Each node represents a job; the **START** node represents the beginning of the workflow. Each line represents a piece of logic. The different options will be broken up into unique groups. GUI fields will be represented with **Name**. If the module and role fields are anything other than lowercase, they will be represented with (name). The logic in workflows is as follows:

- **Green line**: This line of logic is triggered by the success of the previous job.

- **Blue line**: This line of logic is always followed.

- **Red line**: This line of logic is triggered by the failure of the previous job.

- **Convergence** (all_parents_must_converge): The default convergence is that if **ANY** of the logic leads to the node, it is executed. The **ALL** option requires that all of the logic from the parent nodes has to have been met. This means that any success or failure links need to have led to the node.

Each of the following options depends on whether the job template has been set to allow a prompt on launch for the specific option. These options are used within the creation of a workflow node:

- **Node Alias** (`identifier`): The alias to be displayed to the user in the. If using the modules and roles, this must be unique.

- **Node type** (`unified_job_template`): The inventory source, project, job template, workflow job template, or management job name to be used in the node.

- **Inventory**: The inventory to use.

- **Job Type** (`job_type`): The job type to use, either in **Run** or **Check** mode.

- **Limit**: The host pattern to apply.

- **Source Control Branch** (`scm_branch`): The source control branch to use.

- **Verbosity**: The verbosity level to use.

- **Show changes** (`diff_mode`): Shows the changes made by an Ansible task if supported.

- **Job Tags** (`job_tags`): Job tags to use.

- **Skip Tags** (`skip_tags`): Job tags to skip.

- **Survey Questions** (`extra_data`): Survey answers to use in this workflow. Merged with variables when using the API directly, using modules, or using roles.

- **Variables** (`extra_data`): Any extra variables or survey answers to use.

Workflow node lists can be created in the GUI using the following steps:

1. Navigate to **Templates** on the left-hand sidebar.

2. Click on a workflow to edit or change a current workflow node list.

3. Select **Visualizer** from the tabs at the top of the workflow.

4. Click **Start** to begin, or a + icon on a link or node to add another node.

5. Select **On Success** | **On Failure** | **Always** if prompted, and then fill in the form with the fields given in this section.

6. Click **Save**.

7. Repeat as needed to add more job templates or nodes.

Using the GUI to create workflows is doable, but because of how complicated and detailed they can get, they are hard to reproduce. It is recommended to use roles and modules to maintain workflows. However, a different method is used to create the logical flow, which we will discuss in the following section.

Workflow job template options when using modules and roles

The workflow node list describes the logical flow carried out using a list of modules and roles. These use the same options as the GUI but in a more data-centric format.

Each list item can contain options such as extra_data and job_tags, as listed in the previous section, but can also use additional fields. Using the workflows/ set_workflow_using_ module.yml file for reference, let's observe the following:

```
workflow_nodes:
  - identifier: Inventory Update
    related:
    unified_job_template:
    all_parents_must_converge: false
    extra_data: {}
```

The identifier must be unique since it is needed for any lookup done to find the node in the workflow in the future. Additional fields can be added, such as extra_data and all_parents_must_ converge to set the options. The unified_job_template field and the related field are used to set the job to run and create the links. The related field can contain four options:

```
related:
  always_nodes: []
  credentials: []
  failure_nodes: []
  success_nodes:
    - identifier: Template 1
```

The three fields, always_nodes, success_nodes, and failure_nodes, are lists of identifiers to create links in the workflow. The credentials field is a list of credential names to use for that specific node as launch-time prompts. If the list is empty, it does not need to be included.

unified_job_template are not uniform in their fields. There are a few iterations to go over to describe them. Each contains a name field and a type field.

Each type of unified_job_template follows the following form as a base:

```
unified_job_template:
  name: Cleanup Activity Stream
  type: system_job
```

If we know that `unified_job_template` is uniquely named, an organization does not need to be set. In this case, where the `type` is `system_job`, they are built-in job templates that can be launched, for example, if the Automation controller job history needs to be cleaned up, among other tasks.

> **Important Note**
>
> It is highly recommended to include organization fields in `unified_job_template` whenever possible. The object names inside of an organization need to be unique, but users do not always see what other organizations name their job templates or projects, which can lead to errors when creating a workflow.

The `inventory_source` type required its organization to be defined under the `inventory` dictionary:

```
unified_job_template:
  name: RHVM-01
  type: inventory_source
  inventory:
    organization:
      name: Default
```

Specifically, the `inventory.organization.name` can be set to ensure that the correct inventory source is used.

The approval node doesn't rely on an actual unified job template, so it takes on a different form. The type for this is `workflow_approval`:

```
unified_job_template:
  name: Approval to continue
  type: workflow_approval
  description: Approval node for example
  timeout: 900
```

The name can be anything, as can the description. `timeout` sets how long it takes without input from the user before the node fails.

The `project` type, the `job_template` type, and the `workflow_job_template` type share the same form. Workflow nodes can launch other workflows, just as they can launch job templates:

```
unified_job_template:
  name: Test Project
```

```
    type: project
    organization:
      name: Default
```

`organization.name` can be set to ensure that the correct one is used; this is useful if, for example, the same project name is used in multiple organizations. If an organization lookup is not needed, do not include the `organization` dictionary.

Putting everything together for the workflow node should look as follows:

```
  - identifier: Inventory Update
    related:
      credentials: []
      success_nodes:
        - identifier: Template 1
    unified_job_template:
      name: RHVM-01
      inventory:
        organization:
          name: Default
      type: inventory_source
    all_parents_must_converge: false
    extra_data: {}
```

> **Important Note**
>
> The identifier is exactly what it says on the tin: it identifies a particular node in a workflow and should be unique within the workflow. Everything else in the workflow node definition can change, and the module or role will still push the right information to that identifier. While the field is used to display an alias in the Workflow **Visualizer**, it is important to remember it is an identifier.
>
> If removing nodes, in the process of changing identifiers, it is important to either set the `state` option for that node to `absent` or set the `destroy_current_schema` option to `true`. This is so that the node with that identifier is either deleted, or all the nodes are deleted and recreated from scratch.

Putting these together completes a workflow job template node list. With all the pieces to create workflows, the next step is learning about how to use modules and roles to create them.

Creating a workflow using modules

The module is best for modifying an existing workflow or creating a workflow on demand. It's an unconventional use case to use the module to create on-demand workflows using a job template from user inputs. However, creating a workflow in this way can be useful for spinning up virtual machines in a specific cluster, or site-specific, temporary workflows for the user to run.

The module can be invoked from the following task; it is an excerpt from the `//workflkows/ set_workflow_using_module.yml` file. The full file can be found in this book's repository chapter folder. In addition, the playbook creates the necessary inventory, project, credentials, and job templates to be used in the workflow:

```
- name: Push workflows to controller
  ansible.controller.workflow_job_template:
    name: Complicated workflow schema
```

The schema here only contains a single node; however, one for each type of node is included in the aforementioned full playbook file, which covers all the different kinds of nodes:

```
schema:
  - identifier: Inventory Update
    related:
      always_nodes: []
      credentials: []
      failure_nodes: []
      success_nodes:
        - identifier: Template 1
    unified_job_template:
      name: RHVM-01
      inventory:
        organization:
          name: Default
      type: inventory_source
    all_parents_must_converge: false
    extra_data: {}
```

While this module is not one often used on its own when using CaC, it is the basis for using the role to create them, so it is important to know how it works.

> **Fun Fact**
>
> Because it took so long to loop over the `workflow_job_template_node` module to create individual nodes, I spent a lot of time figuring out how to create and link nodes in Python in order to improve the module and improve the node definition system as a result. This was inspired by the original module that was able to do it with a file, using a mapped-out workflow schema that had since broken.

Creating workflows using roles

We will now explore using roles to maintain workflows with CaC. The following code for creating and maintaining workflows with roles is based on the previous section and puts each type of node together to create a workflow. The workflow role takes a list of workflows with options and applies them to the Automation controller:

```
//workflows/set_workflows_with_roles.yml
    controller_workflows:
      - name: Complicated workflow schema
        destroy_current_schema: true
```

Remember that the `related` field comes from the export:

```
            related:
              workflow_nodes:
                - identifier: Template 1
                  all_parents_must_converge: true
                  related:
                    success_nodes:
                      - identifier: Template 2
                    failure_nodes:
                      - identifier: Approval Node
                  unified_job_template:
                    name: test-template-1
                    organization:
                      name: Default
                    type: job_template
```

Also, note that the approval node is used to require user input to move forward with a workflow:

```
                - identifier: Approval Node
                  related:
```

```
                    success_nodes:
                      - identifier: Template 2
                  unified_job_template:
                    description: Approval node for example
                    timeout: 900
                    type: workflow_approval
                    name: Approval to continue
```

The recommended list of roles to use to make sure that all the pieces are in place is as follows:

```
    - redhat_cop.controller_configuration.credentials
    - redhat_cop.controller_configuration.inventories
    - redhat_cop.controller_configuration.inventory_sources
    - redhat_cop.controller_configuration.projects
    - redhat_cop.controller_configuration.job_templates
    - redhat_cop.controller_configuration.workflow_job_
  templates
```

The file example also includes the configuration pieces for the previous code to be invoked alongside the workflow. Workflows allow for logic to flow between job templates, using a structured flow. However, determining how jobs run requires one last piece of the puzzle. Instance groups are used to determine where a job actually runs. Instance groups are discussed in *Chapter 14*, *Automating at Scale with Automation Mesh*.

Workflows and job templates have been discussed, but there is also another tool that can be used. Job slicing allows us to split a job into multiple pieces and we will explore this in the following section.

Using job slicing to slice a job template into multiple jobs

In the *Job template options for the GUI, modules, and roles* section, a particular option came up for `job_slice_count`. This specific feature takes a regular job template and creates a brand-new workflow with the jobs that equal the number of job slices. We can see an example of this in *Figure 10.8*, which shows how a job template with a `job_slice_count` value of 3 is split into three jobs:

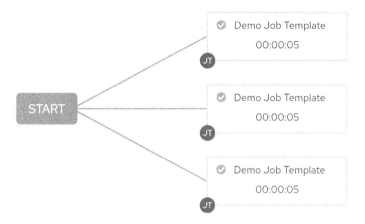

Figure 10.8 – A sliced job template

This is useful for jobs that act on a large number of hosts, as it splits them up among execution nodes. Depending on the number of nodes available, the number of hosts, and the tasks performed in the job, this can decrease the amount of time it takes to execute a job and make it more efficient.

Using a large number of slices is not recommended, as it can adversely affect the job scheduler. The recommended number of slices to use is either equal to or less than the number of execution nodes available. This allows the Automation controller to spread out the workload among the nodes and adjust itself as needed, depending on the resources needed. For example, if there are four controller execution nodes, then the recommended ideal number of job slices is four. This can be adjusted up or down depending on preferences, but the absolute maximum to never exceed is twice the number of execution nodes. While the Automation controller will not limit the number of slices, as more slices are added, a limit on the theoretical return exists.

The one downside to using job slicing is when a sliced job template is used in a workflow. Normally, artifacts are passed between workflow nodes. However, when job slices split inventories, these artifacts are not picked up by the next workflow job. This is because the artifacts differ between the sliced jobs and the merged data can be inconsistent. By looking up the job data using the API, the data can be retrieved. However, in this case, the best option is not to slice jobs that create artifacts for use in workflows.

Summary

This chapter covered projects, job templates, and workflows. While previous chapters have covered many of the different parts of the Automation controller, each of these has led us to the job templates and workflows. This chapter discussed the creation of projects, job templates, and workflows, the use of surveys for job templates and workflows, and the use of job slicing.

The next chapter will go into further detail on job templates and workflows, how to devise a strategy to use them effectively, and the different bells and whistles that we can use to make them more efficient.

Part 3: Extending Ansible Tower

Now that all the pieces have been created, how do you maintain them? How does the Automation controller integrate with other services such as CI/CD pipelines and workflows? How do you connect to other services and scale your automation? These chapters will go over all of those details about how to extend Ansible Automation Platform.

The following chapters are included in this section:

- *Chapter 11, Creating Advanced Workflows and Jobs*
- *Chapter 12, Using CI/CD to Interact with the Automation Controller*
- *Chapter 13, Integration with Other Services*
- *Chapter 14, Automating at Scale with Automation Mesh*
- *Chapter 15, Using Automation Services Catalog*

Creating Advanced Workflows and Jobs

The previous chapter covered creating job templates and workflows inside the Automation controller. This chapter will go into detail about designing playbooks and jobs in a workflow to take advantage of workflows. This includes creating nodes that contain information so that users do not need to hunt in a playbook and using approval nodes to gain user input to allow a workflow to continue. Notifications are used to notify users and services about the specific events that have occurred.

In this chapter, we will cover the following topics:

- Creating advanced workflows
- Using and configuring notifications

Technical requirements

All the code referenced in this chapter is available at `https://github.com/PacktPublishing/Demystifying-Ansible-Automation-Platform/tree/main/ch11`. It is assumed that you have Ansible installed to run the code provided.

Creating advanced workflows

Most users who have been using Ansible with the command line design their playbooks and roles to encompass everything for a given scenario in a single playbook, These monolithic playbooks account for various scenarios in a single play, such as deploying a high-availability SQL server with a load balancer on multiple servers from scratch. However, when using workflows, it makes more sense to break up the work into various parts that are run separately. The advantages of doing this are that you can anticipate and account for failures, pass information from one job node to another, and wait for user approval to continue during the workflow execution.

Using workflow artifacts and variables

Workflows can take advantage of a special set of extra variables that are passed from one node to another. This can be achieved by using a specific Ansible module: `set_stats`. These can be created with the following playbook task:

```
//review_results.yaml
  - set_stats:
      data:
        list_to_pass: "{{ list_to_pass }}"
        host_groups: "{{ group_list }}"
      aggregate: false
      per_host: false
```

The `set_stats` module has the following options:

- `data`: A dictionary of variables to be set and their values.
- `aggregate`: Whether the existing `set_stat` values will be kept. This defaults to yes to add and aggregates to existing stats.
- `per_host`: Whether stats are kept per host or not.

Variables set with `set_stats` are inherited by any job in the workflow that occurs after that job is finished. An example workflow can be seen in the following diagram:

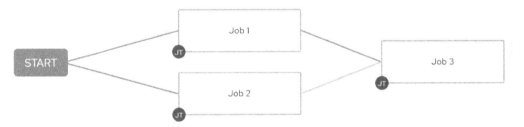

Figure 11.1 – Workflow illustration for converging nodes

Any variables set with `set_stats` in *Job 1* or *Job 2* will be inherited in *Job 3* and any job after that. However, in this case, the merger of *Job 1* and *Job 2* is not controlled or defined, so be sure to set variables to unique names so that there is no overlap. The order of precedence for variables in jobs and workflows is dictated by the following list. The further you must go down the list, the higher the priority for precedence:

- Job template extra variables
- Job template survey

- Job launch extra variables

- Job artifacts

- Workflow job template extra variables

- Workflow job template survey

- Workflow job launch extra variables

It is important to keep this variable precedence in mind when designing workflows, job templates, and Ansible playbooks to use in the Automation controller. The official list regarding variable precedence can be found here: `https://docs.ansible.com/automation-controller/latest/html/userguide/job_templates.html#ug-jobtemplates-extravars`.

In addition, if you want to set a single `set_stats` variable that combines variables from each host, the following task can be used:

```
//set_vars.yaml
    - name: "set a stat that has variables per host"
      set_stats:
        data:
          var_per_host: "{{ var_per_host | default({}) |
combine({inventory_hostname: show_vars }) }}"
```

This will take a previous `set_stats` variable that was created for each host, and then save it for a future job node to use. An example of the result is as follows:

```
var_per_host:
  hostname1:
    other_info: 1
  hostname2:
    other_info: 2
  hostname3:
    other_info: 3
```

The `var_per_host` variable can then be used in a later playbook/job template using the following variable reference:

```
"{{var_per_host[inventory_hostname]}}"
```

Using `set_stats` in this manner takes advantage of job artifacts so that you can pass variables to future job nodes and playbooks. The variables that are used can be large dictionaries that are kept, such as data on an entire router configuration. However, with automation, you may need someone to review what is being done before pushing out the change. One part of this is creating a job node that distills information that users need to review.

Creating nodes for user review

With some workflows where a lot of information is being parsed and a review is needed before final approval, it is best to set up a simple playbook that displays information to the user. This can be done with either a set_stats or debug task in the playbook.

The debug task is normally used for troubleshooting, but it does display a variable in full in the job's output:

```
//set_vars.yaml
    - debug:
        var: var_name
```

This will appear as follows in the job's output:

```
TASK [debug] ****************************
ok: [PSQL2] => {
    "show_vars": {
        "other_info": 1
    }
}
ok: [PSQL3] => {
    "show_vars": {
        "other_info": 1
    }
}
ok: [PSQL1] => {
    "show_vars": {
        "other_info": 1
    }
}
```

Figure 11.2 – Debug output

With set_stats, especially if the previous node has created many artifacts, it might be a good idea to add an underscore to the variable. This makes it go to the top of the list so that a user can easily find it:

```
//set_vars.yaml
    - name: "set a stat that has variables per host"
      set_stats:
```

```
data:
  _var_per_host: "{{ var_per_host }}"
```

This node will show up at the top of the artifacts window on the Automation controller, as follows:

```
1 _var_per_host:
2   PSQL2:
3     other_info: 1
4   PSQL3:
5     other_info: 1
6   PSQL1:
7     other_info: 1
  abc_var: nonenonenone
8
```

Figure 11.3 – Workflow artifacts with underscores go to the top

As mentioned previously, these are useful when using approval nodes, as covered in the next section.

Using workflow approval nodes

Approval nodes are useful for when a manager, another team, or someone else needs to review information from a playbook before moving forward with a workflow. Approval nodes were covered in *Chapter 10, Creating Job Templates and Workflows*. The following is the **Configuration as Code** (**CAC**) definition of a workflow approval node in a workflow:

```
- identifier: Approval Node
  unified_job_template:
    description: Approval node for example
    timeout: 900
    type: workflow_approval
    name: Approval to continue
```

These approval nodes pause a workflow until either they time out or a user approves or denies it. If workflows and job templates are not set to run concurrently, this will cause a bottleneck, preventing other jobs from stalling. This can lead to problems if users do not respond to the approval notifications. This will be discussed in the *Notifications and how to integrate them* section.

If a user denies an approval node, it will mark the workflow as failed. The node behaves as if the approval job failed in terms of logic. If this is the last node in the logical flow, then the workflow will be marked as failed. To avoid the workflow being marked as failed, it might be useful to set the workflow's state as successful after a denied approval.

Marking a workflow as failed or successful

In some cases, some conditions result in failed jobs, which then cause a workflow to fail. This can be due to a variety of things, such as certain checks failing or the user deciding to deny the approval node. Another case would be when a log aggregation service checks the status of the workflows and reports the number that failed throughout the day. In all of these cases, it can be useful to incorporate jobs with tasks that dictate the flow of a workflow. Two playbooks are used in such a case: a success playbook and a fail playbook.

The following is a simple success playbook, though it really could be any valid playbook:

```
//success_playbook.yaml
    - debug:
        msg: Approval Denied, Marking workflow as Successful.
```

An example of when to use this is when designing workflows for checks and work to be done without any changes, then have a user review the results before approving or denying the node to continue and finish the work. If the user chooses to deny, the workflow will be marked as failed, which doesn't look good on reports when this is done hundreds of times a day. This is why the workflow must be marked as a success since it did what it intended to do.

The fail playbook, however, does rely on the `fail` module to make sure the playbook fails:

```
//fail_playbook.yaml
    - name: "Mark workflow/template as Failed"
      fail:
        msg: "Set Status to Fail because an error occurred in
the workflow/template."
```

While these are two simple playbooks, they help determine the outcomes of the workflows when there has been an acceptable failure. They can also be used to force a workflow failure, depending on logic, and trigger a notification. The next section will go into using notifications so that a message can be sent if an event occurs.

Using and configuring notifications

Notifications are ways to send messages somewhere when an event occurs in the Automation controller. These notifications are made through a variety of mediums, such as emails, slack messages, or webhooks.

Understanding the basics of notifications

These are extremely useful for approvals so that users respond to them, but also for critical pieces such as an inventory failure. The important part is that these are knobs that can be turned on and off as needed.

The events that can trigger a notification are as follows:

- **Workflow Approvals**: A workflow approval node is active and needs to be approved or denied.
- **Start**: When a job or a project/inventory synchronization starts.
- **Success**: A notification is sent on success.
- **Failure**: A notification is sent on error.

Each of these sets a state for when to send a notification. However, these events only pertain to some things in the Automation controller. The following objects in the Automation controller can have notifications set. All of these contain **Start**, **Success**, and **Failure** options:

- **Inventory Sources**: For inventory synchronization
- **Projects**: For project synchronization
- **Job Templates**: For jobs that occur
- **Workflow Job Templates**: For jobs that occur and for workflow approval nodes that trigger
- **Organizations**: To turn any of the aforementioned options on globally for the entire organization

Inventory and project failures can interrupt how all the jobs are run, bringing work to a critical stoppage. It is recommended that failure events be turned on for these. However, different groups communicate in different ways. Some prefer emails or Slack notifications. The following notification services can be used to send a notification:

- Email
- Grafana
- IRC
- Mattermost
- Pagerduty
- Rocket.chat
- Slack
- Twilio

Now that we've learned what notifications are and where they are used, let's learn how to configure notification templates.

Notification options for the GUI, modules, and roles

The different options will be broken up into three sections: notification options, configuration, and messages. The GUI fields will be represented with **Name**. If the module and role fields are different other than being in lowercase, they will be represented as `name`.

Notification options

For the GUI, roles, and modules, we can use the following options:

- **Name**: The name of the notification.
- **Description**: A description of the notification.
- **Organization**: The organization that the notification belongs to.
- **Type**: `notification_type` – which service the notification uses. This type is chosen from the aforementioned list – for example, email, Grafana, or Slack.
- **Type Details**: `notification_configuration` – the specifics to connect to the service. This varies, depending on the service.
- **Customize messages**: `messages` – any custom messages to be sent when being notified. There are built-in defaults, but it is possible to override these.

Because each of the services has different configuration options, it is best to consult the documentation at `https://docs.ansible.com/automation-controller/latest/html/userguide/notifications.html` for each setting.

This section will not cover every notification service, but it will cover two of the most popular ones: email and Slack.

Email notification configuration

The following are the configuration settings for email. Here, we are using Gmail as an example. The steps for setting up an app password for Google can be found at `https://support.google.com/accounts/answer/185833`.

Each of these settings is straightforward, such as the host, sender, and recipient. This example uses TLS and port 587 to contact the email server. The full configuration is as follows:

```
//notifications/set_notifications_with_roles.yaml
        notification_configuration:
            username: asdf@gmail.com
```

```
        password: tnbfksktebythfwv
        host: smtp.gmail.com
        recipients:
          - asdf@gmail.com
        sender: asdf@gmail.com
        port: 587
        timeout: 60
        use_tls: true
```

Another popular option for notifications is Slack; the next section will cover how to configure Slack notifications.

Slack notification configuration

As for Slack, there are only a few options. However, it does require a Slack application to be created and a token for said app. The links for app creation can be found in the Automation controller documentation linked at the beginning of the *Notification options* section. The following configuration options must be used:

```
 //notifications/set_notifications_with_roles.yaml
        notification_configuration:
          channels:
            - group_approval_notification
          token: "{{ notification_slack_oath }}"
          use_ssl: false
          use_tls: false
```

Now that we've learned how to configure both email and Slack, the next section will focus on customized notification messages. Some services will want a JSON object to be sent, while others will allow custom text to be sent to users. Customizing notification messages is important so that the right information gets sent.

Notification messages

Each notification type can send custom messages. For email, this includes a subject that has been categorized as a message, and a body – that is, the actual body sent in the mail. Slack, on the other hand, only supports custom messages; there is no body. These fall under the following areas. These are also nested dictionaries that should be used in the modules and roles:

```
 messages
     started
```

```
success
error
workflow_approval
    approved
    denied
    running
    timed_out
```

Each of these contains the message or body to send when the appropriate notification is triggered. A message can be crafted to be sent. The following message has been crafted for Slack:

```
//set_notifications_with_roles.yaml
messages:
  workflow_approval:
    running:
      body: ""
      message: 'The approval node "{  { approval_node_name }}"
needs review. This node can be viewed at: {  { workflow_url }},
job data: {  { job_metadata }}'
```

> **Note**
>
> The double spaces between the curly braces, {   {, prevent Ansible from interpreting the variable. The roles are built to replace this using a `regex_replace` filter. So, the correct information is sent to the Automation controller. It is difficult to send the right text for variables using the module or URI without using a filter.

The body field is not used as Slack does not use the `body` field. The variables that can be used in these messages are limited. They include the following:

- `job`: dict with many suboptions to be used, such as `job.id` and `job.summary_fields. created_by.username`.
- `job_friendly_name`: The name of the job template.
- `url`: The URL of the job, such as `https://controller.node/#/jobs/ playbook/197`.
- `approval_status`: The status of an approval node, such as `approved`.
- `approval_node_name`: The name of the approval node.

- `workflow_url`: The workflow URL, such as `https://controller.node/#/jobs/workflow/127`.

- `job_metadata`: A dictionary of metadata about a job, such as its ID, name, and URL. The suboptions cannot be referenced in the body or message.

Each of these variables can be used to craft custom messages, but the messages are limited to text and these variables. To send an even more customized message, a playbook and task would need to be used, such as to send an email with an attachment. Some additional tweaking can be done to limit the number of approval messages that are sent.

Preventing workflow approval notification messages

While each of the aforementioned options can be turned on/off for messages to be sent, such as success or error, for workflow approval, it is not possible to tweak the suboptions in that way. This means that if the workflow approval notifications are turned on, an approved, denied, running, or timed out message will always be sent. Here, the workaround is to use a message body that contains no data as it will not send a blank message:

```
//notifications/set_notifications_with_roles.yaml
          workflow_approval:
            approved:
              message: "{  { job.name }}"
```

By using the `job.name` variable, which does not exist for an approval node, the message is never sent. This allows each option to be tweaked to dictate the necessary behavior.

Creating a notification template using modules

A module can be invoked from a task. The following is an excerpt from the `//notifications/set_notifications_using_module.yml` file. The full file can be found in this book's GitHub repository:

```
  - name: Add Slack notification with custom messages
    ansible.controller.notification_template:
```

The first section creates the necessary notification:

```
        name: Slack_approval_notification
        organization: Default
        notification_type: slack
```

The second section creates the necessary configuration for email:

```
notification_configuration:
  channels:
    - notification_test
  token: xoxb-1234
```

The last section sets any custom messages to be sent. This has been left blank so that default messages are used:

```
messages:
  started:
    message: "{{ '{{ job_friendly_name }}{{ job.id }}
 started' }}"
```

When considering whether you wish to use a GUI, module, or role, it's the module that doesn't always get used. However, it is important to know how it works. When maintaining and updating notification templates, it makes more sense to make use of roles.

Creating a notification template using roles

Let's learn how to create and maintain job templates with roles. The job templates role takes a list of job templates with options and applies them to the Automation controller:

```
//notifications/set_notifications_with_roles.yml
  controller_notifications:
```

The first section creates the necessary notification:

```
- name: Gmail notification
  description: Notify us on Google
  organization: Default
  notification_type: email
```

The second section creates the necessary configuration for email:

```
notification_configuration:
  username: asdf@gmail.com
  password: tnbfksktebythfwv
  host: smtp.gmail.com
  recipients:
  - asdf@gmail.com
```

```
        sender: asdf@gmail.com
        port: 587
        timeout: 60
        use_tls: true
```

The last section sets any custom messages to be sent. This has been left blank so that default messages are used:

```
        messages:
          success:
            body: '{"fields": {"project": {"id":
  "11111"},"summary": "Lab {  { job.status
              }} Ansible Controller {  { job.name
  }}","description": "{  { job.status }} in {  {
              job.name }} {  { job.id }} {  { url
  }}","issuetype": {"id": "1"}}}'
```

The notification role allows users to maintain the Automation controller's configuration as code.

The `redhat_cop.controller_configuration.notification_templates` role was used in the preceding playbook to push the configuration to the Automation controller.

Adding notifications to activities

Which notification service to use depends on the group preferences of those who want to receive the notifications. It is possible to have notifications sent to *any* combination of events and services, depending on the preference. Depending on the configuration, a valid configuration could include a Slack channel and multiple email notifications, each with a separate list of email recipients. These configuration options can be seen in the workflow notification settings tab:

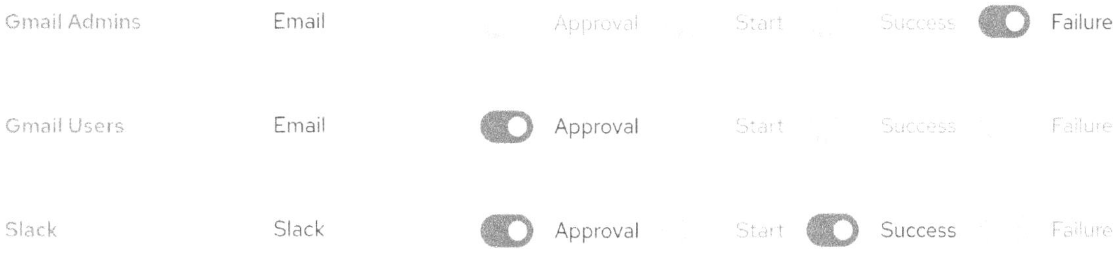

Gmail Admins	Email		Approval	Start	Success	⬤	Failure
Gmail Users	Email	⬤ Approval		Start	Success		Failure
Slack	Slack	⬤ Approval		Start	⬤ Success		Failure

Figure 11.4 – Notifications settings for a workflow

As mentioned in the *Using and configuring notifications* section, notifications can be added to inventory sources, projects, job templates, and workflow templates. In the GUI, it is as simple as toggling the option to turn it on, as illustrated in the preceding screenshot. In the modules and roles, each of the previously mentioned objects accepts a list of notifications to turn on. An example of this can be found in both playbooks in the `//notifications` folder for this chapter. For example, to replicate the notifications shown in the preceding screenshot, you could use the following dictionaries with list items:

```
notification_templates_error:
  - Gmail notification
notification_templates_started: []
notification_templates_success:
  - Slack_notification
notification_templates_approvals:
  - Gmail notification
  - Slack_notification
```

This can be applied to anything that uses a notification template. As mentioned previously, it may make sense to make a few different templates for the same service. While there is no limit to the number of notifications that can be made, there is a limit to users' patience, so use them wisely.

Summary

This chapter covered advanced workflow options and notifications. Some of these will not be applicable in every situation, but they are useful tools. If there is a workflow that takes over 10 minutes or even an hour to complete, some users may prefer a way to get notified of the need to intervene with an approval node or a job failure. The logic that's used in workflows can take advantage of artifacts and other methods to become as simple or as complex as needed.

The next chapter will cover the use of CI/CD and ways to interact with the Automation controller using outside tasks and services.

12

Using CI/CD to Interact with Automation Controller

Previous chapters have gone into detail about using roles and modules to interact with parts of the **Ansible Automation Platform** (**AAP**). This chapter will go into detail about using principles of **Configuration as Code** (**CaC**) and **Continuous Integration and Continuous Delivery** (**CI/CD**) to maintain configuration and interact with services in the AAP. Some of this has been covered in other chapters, such as triggering a project update, a configuration change, when a pull request has been completed, or doing a regular backup of the installation. The goal of this chapter is to put those ideas together in a more cohesive format with examples of how to integrate those examples into CI/CD.

In this chapter, we're going to cover the following main topics:

- A brief introduction to CI/CD pipelines and webhooks
- Maintaining Automation controller and hub through infrastructure as code paired with CI/CD
- Launching jobs, monitoring, and interacting with workflows using CI/CD playbooks
- Ad hoc commands
- Backup and restore options using CI/CD playbook

Technical requirements

This chapter will have multiple playbooks. All code referenced in this chapter is available at https://github.com/PacktPublishing/Demystifying-Ansible-Automation-Platform/tree/main/ch12. It is assumed that you have Ansible installed in order to run the code provided.

A brief introduction to CI/CD pipelines and webhooks

CI/CD pipelines come in many forms. A few examples are GitHub Actions, GitLab Pipelines, Bitbucket Pipelines, and Azure Pipelines. While each has different ways of doing things, they can be summed up as code that is triggered when an event happens from either an event in a repository or an outside request.

Triggers can include merge requests, merges, curl requests, webhooks, or other events. They also can be run on generic or special containers. In many cases, the same container image used for execution environments can be used as a CI/CD runner image.

Refer to the documentation on the Git server that is being used on how to create pipelines specific to that technology. While the actual code implementation will be the same, there can be some key differences. This chapter will go into some detail on GitHub and GitLab implementations.

A webhook is used when an event is set to trigger to reach out to one of the AAP services, usually the Automation controller. This triggers a job and then sends information back. Just like the pipelines, this is unique for each Git service. For webhooks specifically, refer to the webhook documentation for the Automation controller at `https://docs.ansible.com/automation-controller/latest/html/userguide/webhooks.html`.

The limitation is that webhooks can only trigger workflows and job templates. However, that limitation can be eliminated by using modules and roles in the playbook triggered to do whatever actions are needed. An example of this is if you need to use some prompt user on launch variables that do not match up 1:1 in the payload from the Git server. A job can take the payload and then launch another job using the correct prompts.

Another issue with both the webhooks and CI/CD pipelines is that the Git server must be able to reach port `443` of the Automation controller. In most cases, this is not an issue; however, if you're using a public repository on `https://github.com/`, it may not be feasible to reach an internal server.

All of these are tools that can be used to trigger bits of code for any purpose. The following few sections will go over various purposes to use pipelines and webhooks and useful code to use for these actions.

Maintaining an Automation controller and hub through infrastructure as code paired with CI/CD

Keeping projects and other objects updated in the AAP can be painful and burdensome without using automation. CI/CD automation is a great way to solve these problems. This section will focus on using tasks to solve this problem.

There are files that have been introduced, scattered throughout the previous chapters, that set the contents of both Automation hub and the Automation controller in configuration files. When referenced, these files can be used by the `redhat_cop.controller_configuration` roles to manage the Automation controller. These have been collected in the `ch12/controller/configs` folder.

There are two main reasons to trigger code for maintaining state: to test a pull request, or to update a project or a configuration change after a merge has occurred. While the latter maintains state, the first allows for testing and provides checks prior to a merge, as follows:

```
//controller/github_ci.yml
---
```

The first step determines when the code runs. In this case, it runs when a push or merge is made to the main branch as follows:

```
on:
  push:
    branches:
      - main
jobs:
  Controller-configuration:
    name: Deploy configuration to controller
    runs-on: ubuntu-latest
    steps:
```

A job is defined, what container it runs on, and then what actions are taken. While this is an excerpt, steps to install Python and Galaxy requirements were done beforehand. This action runs the playbook with extra variables as inputs, as follows:

```
    - name: "Perform playbook update"
      run: ansible-playbook configure_controller.yml -e
controller_hostname=https://controller.node -e controller_
password=${{ secrets.controller_password }}
```

The same can be done on GitLab using the following code:

```
//controller/.gitlab-ci.yml
---
update_on_mr:
  stage: test
  only:
    - merge_requests
  script:
    - ansible-playbook configure_controller.yaml --extra-vars
"controller_branch=$CI_COMMIT_BRANCH"
```

The key concepts are setting up the environment to run the tasks on, when to trigger a task, and finally the task itself. The environment can be set up beforehand by building purpose-built containers, or through tasks that execute every time. The *when* is determined by events and triggers, unique to the Git system, and finally, the tasks are generally the playbooks to run.

Another playbook to run would be one to update a project. This is key as when a repository merge is completed, that change can then be pushed to the Automation controller as follows:

```
//controller/project_update.yml
---
    - name: Update Project
      ansible.controller.project_update:
        name: Network Playbooks
        organization: Default
        wait: True
        timeout: 600
        interval: 10
```

This can also be applied to push updates for collections and execution environments to the Automation hub when changes and releases occur in their respective Git repositories. An example of this is to publish a built collection to an automation hub as follows:

```
//hub/publish_collection.yml
---
    - name: Publish Collection
      redhat_cop.ah_configuration.ah_collection:
        namespace: custom_collection_space
        name: custom_collection
        path: custom_collection_custom_collection-1.0.0tar.gz
        auto_approve: false
```

While these are good examples of what can be done to keep configurations up to date, it is also possible to use the CI/CD jobs to do more than just keep configurations up to date. The following section will look at using pipelines to launch jobs.

Launching jobs, monitoring, and interacting with workflows using CI/CD playbooks

Another use case for CI/CD is for launching jobs. Examples of this are scheduled jobs and integration tests. It is good practice to run workflows and jobs with test input and check the results. This can help detect changes that have happened in either the code or the environment.

The playbooks used in this section can be used as jobs in the Automation controller or in their respective Git service pipeline. The idea is to use the GitLab/GitHub workflows or a controller job in order to initiate them.

The playbook is a demonstration of what can be done.

The playbook takes several inputs as follows:

- `workflow_name`: Name of the workflow to launch and monitor.
- `workflow_extra_vars_dict`: Dictionary of extra variables to use when launching the workflow.
- `workflow_node_to_check`: Identifier of the workflow node to wait on till it is finished. The data from the API will be returned from this node as well. This can be used for assertions or other checks.
- `approval_template_name`: Approval node name. The module that uses this variable can either approve or deny an approval node.

In addition, the modules used have several common inputs as follows:

- `workflow_job_id`: The ID of the workflow job; this is returned from the job launch module.
- `timeout`: The timeout to wait on the specific task to finish.
- `interval`: The interval at which to check periodically on the task.

The inputs and variables are used in the `workflows/launch_workflows.yaml` playbook that is built to launch and interact with the workflow created in *Chapter 10, Creating Job Templates and Workflows*.

The first task takes the workflow name and extra variables to launch the workflow. It should return data on the running job. The wait should be `false`, otherwise, it will wait until the workflow finishes, as follows:

```
- name: "Launch Workflow {{ workflow_name }}"
  ansible.controller.workflow_launch:
    workflow_template: "{{ workflow_name }}"
    extra_vars:        "{{ workflow_extra_vars_dict }}"
```

```
    wait: false
  register: workflow_data
```

The second task waits for a specific node to finish to get results from that node. `ignore_errors` is important if the job node is designed to fail. This is because the task will report `failed` if the job is failed, halting the playbook.

```
- name: Wait for a workflow node to finish
  ansible.controller.workflow_node_wait:
    workflow_job_id: "{{ workflow_data.id }}"
    name: "{{ workflow_node_to_check }}"
    timeout: 90
    ignore_errors: true
```

This ignore errors is important to use and should be used only if the job is designed to fail. Refer to *Chapter 11, Creating Advanced Workflows and Jobs*, to read about where you would use jobs designed to fail.

The third task will go and approve the approval node as follows:

```
- name: Wait for approval node to activate and approve
  ansible.controller.workflow_approval:
    workflow_job_id: "{{ workflow_data.id }}"
    name: "{{ approval_template_name }}"
    interval: 10
    timeout: 200
    action: approve
```

The final task waits for the workflow to finish, as follows:

```
- name: Wait for Workflow to finish
  ansible.controller.job_wait:
    job_id:   "{{ workflow_data.id }}"
    job_type: "workflow_jobs"
    interval: 30
    timeout:  1000
```

This is a framework for a job that launches and monitors a workflow. This can be expanded to test for expected outputs or results using the `assert` module. A task that can check specific output against expected output can be formed as follows:

```
- name: assert output matches expected
  assert:
    that:
      - output_from_workflow == expected_ output_from_workflow
```

These modules in the preceding playbook were made for interacting with the Automation controller to control jobs. They are tools for designing custom integration tests depending on what the job template or workflow needs. The following section will cover another tool to use, which is ad hoc commands to run a specific module against a set of hosts.

Ad hoc commands

On most occasions, there is a call to use an actual playbook, but when things start getting to playbooks within playbooks, or maybe information is needed on another host that is not in the inventory, it's possible to get creative. We have never actually found a reason to use the ad hoc modules and roles; however, it is good to know that they are there. Additionally, if there ever was a reason to use them, it would be in a CI/CD job outside of the Automation controller.

The modules have several important inputs as follows:

- `module_name`: The name of the module to use. This is limited, but more modules can be added in the Automation controller settings.
- `module_args`: The arguments to use for the command.
- `inventory`: Inventory to use for the command.
- `credential`: Credential to use for the command.
- `interval`: The interval to use when checking for an update on the job.
- `timeout`: Time to wait on the job to end.
- `wait`: Boolean, whether to wait on the job to finish.

There are other inputs that the module can take, which can be found in the module documentation, but these are the important ones. The playbook is an example of using the ad hoc modules to start the command task, wait for it to finish, and cancel an in-progress ad hoc job with their corresponding modules:

```
//ad_hoc/launch_ad_hoc.yaml
```

The first step is to launch the shell command as follows:

```
- name: Launch an ad hoc command
  ansible.controller.ad_hoc_command:
    module_name: shell
    module_args: ls -a
    inventory: Demo Inventory
    credential: Demo Credential
    wait: false
  register: command
```

The second step is to wait for it to complete as follows:

```
- name: Wait for ad hoc command max 120s
  ansible.controller.ad_hoc_command_wait:
    command_id: "{{ command.id }}"
    timeout: 120
```

The final module allows the command job to be canceled as follows:

```
- name: Wait for ad hoc command max 120s
  ansible.controller.ad_hoc_command_cancel:
    command_id: "{{ command.id }}"
    timeout: 120
```

The modules return data about the job, just like the other job launch modules. A more useful use case is to use CI/CD to run regular backups and a process to restore a backup for the AAP installation, which is covered in the following section.

Backup and restore options using CI/CD playbook

Backup and restore commands should not be run as jobs or templates on the Automation controller they are being used against. This makes them ideal candidates for CI/CD playbooks. These playbooks use the same configuration files from *Chapter 2, Installing Ansible Automation Platform*, and uses them to backup and restore an AAP installation. For demonstration purposes, these files are available in this chapter's repository.

The playbook takes several inputs as follows:

- `aap_setup_down_offline_token`: Token to be used to download the setup file. It can be generated at `https://access.redhat.com/management/api/`.

- `aap_setup_working_dir`: Working directory to download the setup file, create inventory, and do other work.

- `aap_backup_dest`: Destination directory for the backup file.

- `aap_setup_down_version`: Version of AAP to download, currently 2.2. By default, the downloader will grab the latest.

The playbook will load the variable file from the installation that was created in *Chapter 2, Installing Ansible Automation Platform*, and then run through the roles to download, prepare, and back up the installation, as follows:

```
//backup_restore/aap_backup.yaml
---
  vars_files:
    - inventory_vars/variables.yml
  roles:
    - redhat_cop.aap_utilities.aap_setup_download
    - redhat_cop.aap_utilities.aap_setup_prepare
    - redhat_cop.aap_utilities.aap_backup
```

It is recommended to copy the backup from the server the backup is located on to another location. This can be done with the following task:

```
- name: Copy Backup to another directory
  copy:
    src: "{{aap_setup_working_dir}}/backup/automation-platform-backup-latest.tar.gz"
    dest: "automation-platform-backup-{{ ansible_date_time.date }}.tar.gz"
```

For the restore, the variable to add to the mix is `aap_restore_file` to specify the location of the backup. The only difference is instead of running the `aap_backup` role, use the `aap_restore` role. The playbook can be found at `backup_restore/aap_backup.yaml`.

It is recommended to spin up temporary servers and restore the backup files periodically. A backup is never a backup unless it's been tested.

Summary

This chapter has gone over several tools and jobs to use with webhooks or CI/CD pipelines. It has covered CI/CD pipelines, workflow tools, ad hoc commands, and backup and restore options.

The following chapter will go into integrating your Automation controller with other logging and monitoring services.

Further reading

This chapter discussed various CI/CD pipelines. The documentation for some of the most popular pipelines can be found at the following links:

- GitHub: `https://docs.gitlab.com/ee/ci/pipelines/`
- GitLab: `https://docs.github.com/en/actions/learn-github-actions/understanding-github-actions`
- Bitbucket: `https://support.atlassian.com/bitbucket-cloud/docs/configure-bitbucket-pipelinesyml/`
- Azure DevOps: `https://docs.microsoft.com/en-us/azure/devops/pipelines/?view=azure-devops`
- Google Cloud Source Repositories: `https://cloud.google.com/kubernetes-engine/docs/tutorials/gitops-cloud-build`
- AWS CodeCommit: `https://docs.aws.amazon.com/codepipeline/latest/userguide/tutorials-simple-codecommit.html`

13

Integration with Other Services

In the previous chapters, we looked at the different parts of **Ansible Automation Platform** (**AAP**). In this chapter, we will look at using other services and how they interact with some of the lesser-known parts of AAP. These different services each focus on a different aspect of information generated by AAP. This includes job and event logs and metrics. Having a searchable log that contains this information can be invaluable for detecting problems and trends as your automation grows.

In this chapter, we're going to cover the following main topics:

- Logging services
- Automation Analytics for Red Hat AAP
- Prometheus metrics logging

Technical requirements

All the code referenced in this chapter is available at `https://github.com/PacktPublishing/Demystifying-Ansible-Automation-Platform/tree/main/ch13`. It is assumed that you have Ansible installed to run the code provided.

Logging services

Most enterprise deployments of AAP use some form of log aggregation to get a better view of utilization and trends. The AAP services provide good logs and views of general information and tasks, but it can be hard to keep track as things grow. This section will explain how to deploy and integrate the Automation controller with the Splunk service, but the logging aggregation on the controller is also compatible with Loggly, Sumologic, and the Elastic Stack. More details can be found on the documentation site: `https://docs.ansible.com/automation-controller/latest/html/administration/logging.html`.

> **Note**
>
> Numerous times, I've heard people asking, *"Why doesn't AAP have better logging capabilities?"* And the answer I always come back with is that AAP is not a kitchen sink. There are numerous other services out there that will do a better job than the Ansible team would be able to build and support, so they focus on making the product interact as best as possible with external services, rather than building their own.

Creating an event collector on Splunk

Splunk can be downloaded and installed on a variety of servers. More information about how to install and configure a Splunk server can be found at `https://www.splunk.com/`.

The steps for connecting Splunk to an Automation controller are as follows:

1. Log into the Splunk web interface.
2. Click **Add Data | Monitor | HTTP Event Collector**.
3. Fill in the **Name** field and click **Next**.
4. On the **Select Source Type** page, make sure **Automatic** is selected.
5. Click **Create a new index** and give it a name such as `ansible` or `controller`. Then, click **Save**.
6. Click **Review**. You should see the following:

Review

```
Input Type ............................. Token
Name ......................................... controller
Source name override ......... N/A
Description ............................. controller
Enable indexer acknowledg No
Output Group ......................... N/A
Allowed indexes ...................

                                          ansible



Default index ......................... ansible
Source Type ........................... Automatic
App Context ........................... search
```

Figure 13.1 – Splunk Event Collector settings

7. Click **Submit**.

8. Go to **Settings | Data Inputs | HTTP Event Collector**. You should see something like the following:

Name ▲	Actions	Token Value ⇕	Source Type ⇕	Index ⇕	Status ⇕
controller	Edit	9008cca6-157f-49c7-		ansible	Disabled
controller	Delete	be35-9c5771f8d6ef			

Figure 13.2 – Splunk Event Collector status page

9. Navigate to **Global settings** at the top right and **enable** tokens.

10. Navigate to **Settings | Server Settings | General Settings | Set Enable SSL (HTTPS) in Splunk Web** and set it to **On**.

11. Also, make sure that ports 8000 and 8088 are open on the firewall of the Splunk server.

With the connector on Splunk created, the next step is to start sending events from the Automation controller.

Connecting the Automation controller to Splunk

To connect Splunk to the Automation controller, navigate to the Automation controller's **Settings** page. Then, select **Logging settings**.

The following fields need to be changed:

- **Logging Aggregator**: The web address for the logging aggregator – that is, https://10.242.42.36/services/collector/event.

- **Port**: The port to use. By default, this is 8088.

- **Type**: The type of logging aggregator – in this case, Splunk.

- **Username**: The admin. This is not specified in Splunk and there will be errors unless it's set.

- **Password**: This is the token from the HTTP event collector.

- **Individual Facts**: This is a Boolean that's set to true. This allows for better searching as facts are sent individually instead of in a group.

- **Enable External Logging**: This turns the logging on and off.

- **Enable Certificate Verification**: Turn this on or off, depending on whether you want to verify the Splunk server's certificates.

The page should look like this:

Logging Aggregator ⑦	https://10.242.42.36/ services/collector/event	Logging Aggregator Level Threshold ⑦	INFO	Logging Aggregator Password/Token ⑦	[encrypted]
Logging Aggregator Port ⑦	8088	Logging Aggregator Protocol ⑦	https	TCP Connection Timeout ⑦	5 seconds
Logging Aggregator Type ⑦	splunk	Logging Aggregator Username ⑦	admin	Log Format For API 4XX Errors ⑦	status {status_code} received by user {user_name} attempting to access {url_path} from {remote_addr}
Enable External Logging ⑦	On	Log System Tracking Facts Individually ⑦	On	Enable/disable HTTPS certificate verification ⑦	Off

Figure 13.3 – Automation controller logging settings

In addition, the settings to use for roles or modules, as well as a playbook to implement them, can be found at //splunk/splunk_settings.yaml:

```
controller_settings:
  settings:
    LOG_AGGREGATOR_HOST: https://10.242.42.36/services/
collector/event
    LOG_AGGREGATOR_PORT: 8088
    LOG_AGGREGATOR_TYPE: splunk
    LOG_AGGREGATOR_USERNAME: admin
    LOG_AGGREGATOR_PASSWORD: 9008cca6-157f-49c7-be35-
9c5771f8d6ef
    LOG_AGGREGATOR_INDIVIDUAL_FACTS: true
    LOG_AGGREGATOR_ENABLED: true
    LOG_AGGREGATOR_VERIFY_CERT: false
```

This should start the process of sending data to Splunk. If no events appear in the Ansible index, then it is best to look in the logs. Now, let's learn how to troubleshoot issues with the log aggregator.

Troubleshooting with the log aggregator

To troubleshoot any problems with the log aggregator, check /var/log/tower/rsyslog.err on an Automation controller node.

The following are a few of the errors I found while troubleshooting:

- The requested URL was not found on this server.: When HTTPS was not used
- Token is required: When the admin username was not set

The full error outputs can be found at //splunk/splunk_errors.

Splunk search queries

To search the data that's been gathered, navigate to **Search & Reporting** on the Splunk dashboard.

Here, you can use the following commands:

- See user login history:

```
index=ansible message="User * logged *"
```

- Find the most used playbooks:

```
index=ansible | chart count by event_data.playbook | sort
by count
```

- Find the most used modules being run:

```
index=ansible | chart count by event_data.task_action
```

- Find the jobs that hosts have run, removing null data:

```
index=ansible "event_data.remote_addr" !=NULL AND "event_
data.playbook" !=NULL | chart count by event_data.remote_
addr event_data.playbook
```

- See what tasks are using what modules:

```
index=ansible | chart count by event_data.task_action
event_data.task
```

This is not an exhaustive list of searches that can be made, but it is a good representation of what can be done. These searches can then be used to create dashboards.

Splunk dashboards

These search commands can be used in dashboards to create graphs similar to the following:

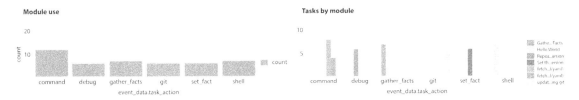

Figure 13.4 – Splunk dashboard graphics

To create a dashboard, follow these steps. This example will use the most used modules search:

1. Navigate to `https://splunk_address/en-US/app/search/dashboards`.

2. Click **Create New Dashboard** and give your dashboard a **Name** and **Description**. Decide whether it should be a **Private** or **Public** dashboard for others to see. Then, click **Dashboard Studio | Grid | Create**.

3. Click the **Chart** icon, as shown in the following screenshot, and choose which type of chart to make:

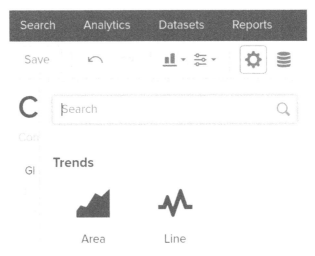

Figure 13.5 – Creating a Splunk dashboard

4. Select **Column**, fill in **Data Source Name**, and fill in **Search with SPL** with `index=ansible | chart count by event_data.task_action`. Click **Run and Save**.

5. The form provides options for the title, axes titles, and positional options – that is, everything to customize the given chart.

6. From here, more charts can be created, or the dashboard can be saved by clicking **Save** in the top-left-hand corner.

Splunk and other log aggregators can be useful tools to supplement the Automation controller and can provide insight into usage and other trends over time. Another tool that can help you do some of this is Red Hat Insights for AAP.

Automation Analytics for Red Hat AAP

Automation Analytics is a tool for exploring the performance and return on investment of either single or multiple Automation controllers. There are also reports, savings calculators, job explorers, and notifications of cluster status. These can be viewed by going to `https://console.redhat.com/ansible`. This section will go through the basic configuration for the Automation controller so that you can send reports to Red Hat, as well as some of the features it provides. The first step is configuring the Automation Analytics reporting.

Automation Analytics configuration

To configure Automation Analytics in the GUI, follow these steps:

1. From the Automation controller, navigate to **Settings | Miscellaneous System settings**. Then, click **Edit**.

    Toggle **Gather data for Automation Analytics** on and fill in the fields for **Red Hat customer username** and **Red Hat customer password**:

Figure 13.6 – Red Hat Insights settings

2. Click **Save**.

3. The **Last gathered entries from the data collection service of Automation Analytics** field should populate shortly after.

4. An update can be forced from the command line by using the sudo `awx-manage gather_analytics --ship` command. The result should look like this:

    ```
    [root@controller username]# awx-manage gather_analytics
    --ship
    /tmp/d72a5fec-5f8e-41a8-8568-f3536906
    77a1-2022-06-11-204019+0000-0.tar.gz
    /tmp/d72a5fec-5f8e-41a8-8568-f3536906
    77a1-2022-06-11-204019+0000-1.tar.gz
    ```

The result should show up on `console.redhat.com` after 5 to 10 minutes, depending on the time needed to process the data. The default time to send new data is 4 hours. This can be changed with the **Automation Analytics Gather Interval** setting.

Configuring the insights settings with roles/modules

The configuration in the preceding GUI can be configured using the settings for roles or modules. A playbook for implementing them can be found at `//analytics/analytics_settings.yaml`:

```
controller_settings:
  settings:
        INSIGHTS_TRACKING_STATE: true
        REDHAT_USERNAME: email@email.com
        REDHAT_PASSWORD: nothing
        AUTOMATION_ANALYTICS_GATHER_INTERVAL: 14400
```

With the integration configured, let's cover what can be done with the reported information.

Charts and other information accessible from the dashboard

Reports and an overview of the Automation controllers are the biggest features that can be gained from the reported information. The following reports can be accessed:

- Jobs/tasks by organization report
- Hosts by organization
- Templates explorer
- Job template run rate
- Changes made by job template
- Hosts changed by job template
- Module usage by organization
- Module usage by job template
- Module usage by task
- Most used modules

In addition, there is also a job explorer that gives a brief rundown of the jobs that have run, similar to the **Jobs** tab on the Automation controller itself. Information about what data is uploaded to generate these reports can be found here: `https://docs.ansible.com/automation-controller/latest/html/administration/usability_data_collection.html#automation-analytics`.

In addition to these reports, there is an Automation calculator.

Automation calculator

There is also an Automation calculator, which allows the user to input the costs of manually doing the changes per hour, the cost of the automation, and set the time it would take to do the jobs manually. This will help generate an estimate of how much time is saved per year by using automation:

Figure 13.7 – Red Hat Insights Automation calculator

The calculator can be configured for each template to reach a rough estimation. It is not a perfect tool, but it is a good approximation for using in reports to justify expenditures.

Beyond the reports, one of the key advantages of Automation Analytics is its ability to observe multiple clusters from a single user interface. Another popular monitoring tool for logging metrics is Prometheus, which we will cover in the next section.

Prometheus metrics logging

Prometheus is a monitoring and observability tool that can incorporate data from the Automation controllers via /api/v2/metrics/. Using the data reported over time, we can gather information about user sessions, the number of jobs running on each node, and license counts. All of the data that's been tracked can be viewed in the metrics endpoint. Now, let's learn how to install and configure Prometheus so that it can be connected to the Automation controller.

Installation and configuration

A playbook has been created that uses the cloudalchemy.prometheus role to install and configure a basic installation of Prometheus on a server. The basic configuration uses the following variables:

- static_configs.targets: The servers to target.
- bearer_token: The OAuth token to use to authenticate. The username or password can also be used.
- metrics_path: The path on the API to read from.

The playbook for installation can be found at //prometheus/prometheus_install.yaml:

```
vars:
  prometheus_scrape_configs:
    - job_name: 'controller.node'
      tls_config:
          insecure_skip_verify: True
      metrics_path: /api/v2/metrics
      scrape_interval: 5s
      scheme: https
      bearer_token: token
      static_configs:
          - targets:
              - controller.node
```

Running this playbook will target a server, install Prometheus, and configure it to connect to the Automation controller. Use the ansible-playbook -i inventory prometheus_install. yaml command to do so.

Prometheus Graphs

Graphs for Prometheus are very simple. You can use a query on the variable from the metrics API, which reports how that value has changed over time. A good example of this is the status total searches:

- `awx_status_total`
- `awx_status_total{status="error"}`
- `awx_status_total{status="failed"}`
- `awx_status_total{status="canceled"}`
- `awx_status_total{status="successful"}`

The `awx_status` primary field's total will return data on all four of the subfields; a search that includes the status will only return that field. This will result in a graph:

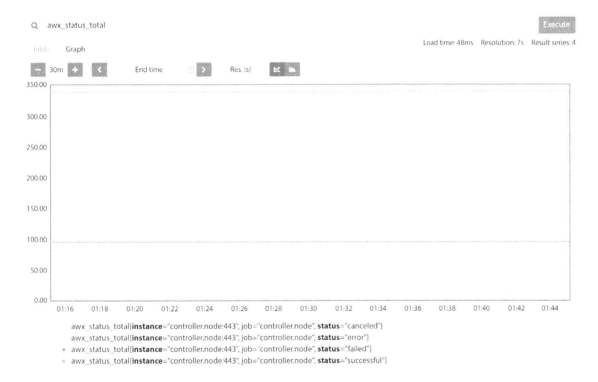

Figure 13.8 – Prometheus awx_status graph

The graph can be customized for a specific period or to show data on other fields. Prometheus Alert Manager can also be configured to send alerts, should certain conditions be met. Overall, it is a useful tool to see historical data and use it for monitoring purposes.

Summary

This chapter covered integration with logging services, Prometheus, and Red Hat Insights. These are useful tools to get more information and reports out of the automation services that are being utilized. While this chapter just provided a brief overview, it only scratches the surface of what can be done with the available information. The next chapter will cover Automation mesh, which allows us to scale the Automation controller using nodes in disparate networks to execute jobs.

14

Automating at Scale with Automation Mesh

Automation mesh is a concept that was released with Ansible Automation Platform 2.1. Its goal was to improve scalability and reliability. This concept allows you to split the workload among nodes between control and execution and to set nodes in a disparate network. Much of this is high-level, so this chapter aims to explain the different types of nodes and their use, and then cover examples to demystify this new feature.

In this chapter, we're going to cover the following main topics:

- Overview of Automation mesh
- Using instance groups with Automation mesh
- Examples of various use cases and how to use Automation mesh

Technical requirements

This chapter will go over Automation mesh nodes. All code referenced in this chapter is available at `https://github.com/PacktPublishing/Demystifying-Ansible-Automation-Platform/tree/main/ch14`. While much of this chapter will cover examples of various deployments, if you wish to follow along, it is suggested that you have several virtual or physical machines ready to be used for deployment. In addition, either an OpenShift or a CodeReady Containers environment will be of use. Please review *Chapter 2, Installing Ansible Automation Platform*, for more details on how to install AAP. While this chapter won't involve installing AAP on OpenShift, it will discuss how to extend a machine installation to take advantage of OpenShift.

Overview of Automation mesh

A new feature called Automation mesh was introduced starting with Ansible Automation Platform 2.1. This allows for the separation of the control plane and the execution parts of the controller. It also allows for execution nodes to be on-premises, in the cloud, or even on the edge.

Increasing the capacity of the Automation controller is done by adding execution nodes or adding CPU and memory to existing nodes. Adding nodes is currently done by adding entries to the inventory and rerunning the installer. On the roadmap of future features is the ability to do this without a full reinstall. The next section will go into detail about the nodes that can be used on AAP.

The distinct types of nodes

There are *seven* types of nodes in an AAP deployment:

- **Control node**: This node contains the web GUI and API and controls all jobs.

- **Execution node**: This node only executes jobs.

- **Hybrid node**: This node combines a control and execution node.

- **Hop node**: Also known as a jump host, this node does not execute jobs but routes traffic to other nodes. It requires minimal ports open for use.

- **Database node**: This node hosts the PostgreSQL database.

- **Automation hub node**: This node hosts the pulp files and the GUI and API.

- **Automation catalog node**: The node hosts the pulp files and the GUI and API.

Each of these can be used in a full installation of AAP, but not all of them are part of Automation mesh.

The following ports need to be opened for services on their respective nodes:

Protocol	Port	Purpose	Nodes
SSH	22/TCP	AAP installation	All
HTTP/HTTPS	80/443/TCP	Web UI, API	Control, hybrid, hub, service catalog
HTTPS	443/TCP	Execution environment (EE) pulls	Execution, hybrid
Postgres	5432/TCP	Database connection	Control, non peered execution*
Receptor	27199/TCP*	Automation mesh	All
HTTP/HTTPS	8080/8443	SSO	SSO

Table 14.1 – Ansible Automation Platform ports required

Hop nodes are specifically designed to require minimal ports open for use. The hop nodes are used to manage or access devices in another security zone. They only require a single port for the receptor to be open for communication.

To better illustrate how the different nodes interact with each other and which ports are in use when communicating, please refer to the following diagram:

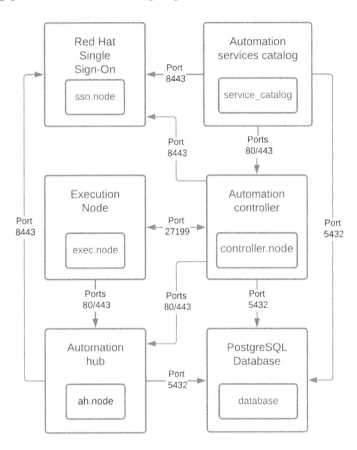

Figure 14.1 – Ansible Automation service connections

In addition, it is important to note that execution nodes need the 80 or 443 port open to pull execution environments from a registry. This is the recommended method of distributing images to execution nodes.

Execution nodes are grouped into what's known as instance groups. Instance groups are logical groups of execution nodes that can be used to specify where job templates or hosts in an inventory can be run. It is important to note that while job templates and inventories can have specified instance groups, they cannot specify an individual node to run on. Each instance group can contain either one or multiple nodes for jobs to execute on.

While many groups can be created, if there are no geographic or network restrictions on the instances, it is best to put them together in the same instance group. Adding additional instance groups works well when there are network and geographic limitations.

For example, if you have an instance group that has instances inside AWS regions for accessing hosts on AWS, having additional instance groups in an enterprise data center may make sense. This would facilitate mitigating different firewall and security concerns that would apply to those connections. However, if the central Automation controller default group can access the inventory hosts that the execution nodes would reach directly, without a large amount of latency, it would be better to utilize the default group. The Automation controller is very good at balancing jobs and workloads between inventories, so the groups and mesh should be utilized specifically to address geographic and network limitations.

There are many reasons to create instance groups to separate where jobs are executed on specific hosts, and they vary use case by use case, but in the end, use your best judgment on how to utilize these features.

The next section will cover a solution to the issue of the execution nodes needing to access container images. The best solution is to allow access to Automation hub, but if that is not feasible, this is a suitable solution.

Distributing execution environments in a restricted network

If there are restrictive firewalls in place, you have two options: having a container registry in the same network as the execution nodes for them to pull from, or pushing images to the remote execution nodes. This can be done by adding them to their execution node internal images. This can be done by following these steps:

1. Use the `podman save image_name` command to save the image to a local file.

2. Copy that local file to the execution node in some manner. This could be `skopeo`, `scp`, saving that file in a project that runs a job, or another manner.

3. On the execution node, become user `awx` and use `podman load image_name` to load the image to the local registry.

4. Make sure that the image registry is `localhost` by using the `podman tag image_name:latest localhost/image_name:latest` command.

5. Edit `/etc/tower/conf.d/execution_environments.py` so that it includes the new image. This should be a new entry in a json list in the following format:

```
{'name': 'Custom EE name', 'image': ' localhost/image_
name:latest '},
```

6. Apply the changes that you made in the file using the `awx-manage register_default_execution_environments` command.

7. Repeat this on all the execution nodes that will use this image.

8. In the Automation controller, set that execution environment to **Never pull container before running**.

This method was partially developed using methods discussed in a Red Hat Knowledge Base article about placing images in a disconnected environment. Please refer to the *Place the execution environment container images* section of this article in case this method changes: `https://access.redhat.com/solutions/6635021`.

This covers the basics of nodes and ports and other considerations to take into account when designing a deployment. The next section will cover grouping execution nodes for use.

Using instance groups with Automation mesh

In addition to instance groups, there are also container groups. These behave like instance groups but are connections to an OpenShift pod that is used to run jobs.

In Automation mesh, each instance is linked together in the mesh using a receptor, allowing multiple paths back to the control nodes. This can be seen in the following diagram:

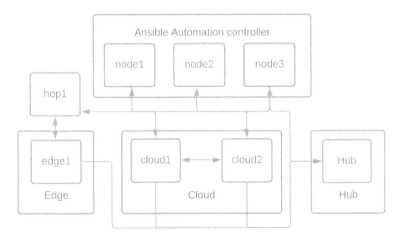

Figure 14.2 – Ansible Automation instance group connections

Instance groups are a vital part of Automation mesh. It is how job templates target which nodes to run jobs on. The next section will cover the different ways to create instance groups.

Creating instance groups

Instance group creation is important as it allows you to bundle instances together for reference. The next few sections will go into the various ways of creating instance groups.

Creating instance groups using the installer

Instance groups can be made when running the installer by putting the node in an Ansible group that begins with `instance_group_`. An example of this can be found in `ch02/inventory_vars/variables.yml`:

```
instance_group_edge:
   - exec
```

This would put the `exec` instance in the edge instance group. This should be the default method of creating instance groups, but they can be reassigned and changed using other methods. The next section will cover creating instance groups in the GUI.

Creating instance groups in the GUI

Instance groups can also be created using the GUI using existing instances. This can be done by following these steps:

1. On the Automation controller, navigate to **Instance Groups | Add | Add instance group**.

2. Set the following fields:

    *   **Name**: The name of the instance group.

    *   **Policy instance minimum**: The minimum number of instances to automatically add to this group when new instances come online. This defaults to `0`.

    *   **Policy instance percentage**: The minimum percentage of instances to automatically add to this group when new instances come online. This defaults to `0`.

    Both of these policy settings can be useful for scaling, but since the installer needs to be run to add new instances, it is best to declare which instance belongs to which group ahead of time.

3. Click **Add**. Navigate to the **Instances** tab and click **Associate**:

Figure 14.3 – Instance group association

4. Select the checkboxes of the instances you wish to add to this instance group. Then, click **Save**.

The other option in the GUI is to create groups as container groups. Let's take a look.

Creating container groups using the GUI

Container groups are created in the same way that instance groups are. However, they have different fields:

- **Name**: The name of the container group.
- **Credential**: The credential to authenticate with OpenShift. This can only be an OpenShift or Kubernetes API Bearer Token type of credential.
- **Pod spec override**: This is a Boolean value that's true if you're using a custom pod specification.
- **Custom pod spec**: This is a custom Kubernetes or OpenShift pod specification. More details about OpenShift pods can be found at `https://docs.openshift.com/online/pro/architecture/core_concepts/pods_and_services.html`.

The `podspec_default file` can be found in `ch14/podspec_default.yml`. The important field to change is the image field. It should be the execution environment image that is to be used in this container group. The default value is `registry.redhat.io/ansible-automation-platform-22/ee-supported-rhel8:latest`. The other fields should be kept as-is unless there is a reason to change them in the OpenShift environment.

These options for instance and container groups can also be created using modules and roles. The next section covers the options for creating instance groups with playbooks.

Instance group settings for modules and roles

The following settings fields are used in both modules and roles to create instance groups:

- `name`: The name of the instance group.
- `new_name`: The new name to rename the named instance group.

- `credential`: The credential to authenticate with OpenShift. This can only be an OpenShift or Kubernetes API Bearer Token type of credential. It is used for container groups.

- `is_container_group`: This is a Boolean value that's true if this is a container group.

- `policy_instance_percentage`: Minimum number of instances to automatically add to this group when new instances come online.

- `policy_instance_minimum`: Static minimum percentage of instances to automatically add to this group when new instances come online.

- `policy_instance_list`: List of exact-match instances that will be assigned to this group.

- `pod_spec_override`: A custom Kubernetes or OpenShift pod specification.

- `instances`: List of instances to add to the instance group.

- `state`: The desired state of the resource, either present or absent.

These fields are mainly used to update the pod spec execution environment or change what nodes are in each instance group. The next section will cover using modules to update instance groups.

Instance group creation using modules

Modules can be used to update the pod spec when it changes or to make quick changes to instances that have been assigned to different instance groups if needed. The following is an excerpt from the playbook using the `instance_group` module:

```
ch14/set_instance_group_using_module.yml
    - name: Add instances to instance group.
      ansible.controller.instance_group:
        name: edge
        instances:
          - exec
```

The better option is to define instance groups in code using roles, which is covered in the next section.

Instance group creation using roles

Roles allow you to define instance groups in code. A role takes a list using the `controller_instance_groups` top-level variable. The following is an excerpt from the playbook using the `instance_groups` role:

```
ch14/set_instance_group_with_roles.yml
    controller_instance_groups:
      - name: edge
```

```
    instances:
      - exec
  - name: cloud
```

The role is then invoked using the `redhat_cop.controller_configuration.instance_groups` role.

After using the installation to set instance groups, you can use **Configuration as Code** (**CaC**) to create instance groups. Instance groups should largely be static, and set in the installer inventory. However, the CaC use of the role can be used to prevent drift, though this should largely be unnecessary. The next section will cover a few examples of using Automation mesh.

Examples of various use cases and how to use Automation mesh

Automation mesh can be used in a variety of ways, but the two key reasons to use it are to remain geographic or due to network restrictions. The following are some examples:

- Security reasons such as a firewalled DMZ
- Separation of private data centers from the public cloud
- Unstable connections that may not survive the duration of a job run
- Remote locations with high latency from the central deployment

Now, let's look at a scenario that uses Automation mesh effectively with these key reasons in mind.

Global Automation mesh example

The following example tries to highlight a setup that covers all the preceding examples. For illustration purposes, each instance group has two nodes. Each group also has at least two servers for redundancy purposes. To start, let's see what how this deployment looks like:

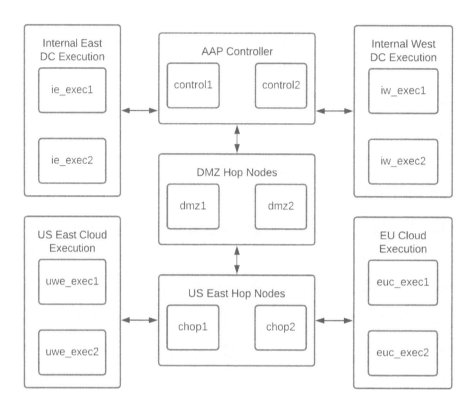

Figure 14.4 – Global Automation mesh example

Using two nodes in each group is only for illustration purposes. In each of the groups, there would be more execution nodes. The central AAP controller group would have three to four servers.

In actual enterprise deployments, each of the execution groups should have two or more nodes. The control nodes at the base of this installation handle the management overhead. This includes project updates, inventory synchronization, management of where jobs are run, the user interface, and the API for this entire deployment.

Please refer to *Chapter 2, Installing Ansible Automation Platform*, about the sizing of the different region's nodes as it is important to monitor each of the nodes to tell when they need more resources, whether the nodes could be scaled down, or whether any scheduled jobs should be spread out to take advantage of off-peak hours.

Now, let's look at the deployment in more detail.

Details of the deployment

This deployment has a central AAP controller group setup with control nodes, and these are directly connected to two data centers on the East and West Coast, respectively. This controller is connected to the company's DMZ. This restricts access to the controller nodes and the data center connections while allowing communications in and out of the receptor ports for the cloud nodes.

The next step is to connect to a pair of hop nodes inside the cloud. This will reduce direct access to the execution nodes and serve as a hub for all the central communication to happen. From there, the cloud execution groups in the US East and EU connect back to the East hop nodes. This is important as connections from the US East have latency times to the EU servers that average around 100 ms. Having high latency between the execution nodes and the targeted hosts makes it so that jobs take longer to execute. Having execution nodes close to the hosts they are targeting eliminates latency and speeds up job execution times. The next section deals with how the mesh peers nodes with each other.

How the mesh peers nodes

The mesh peers nodes in the form of a web. This can be seen in the following diagram:

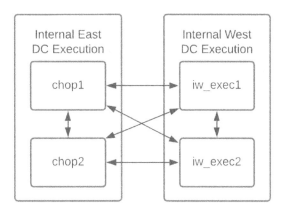

Figure 14.5 – Peered node connections

This illustrates the redundancy built into the mesh and that the nodes are peered to each other and connected. Peers define node-to-node connections and are set at either the host variable level or the group variable level. In this example, the Internal West group is peered to the Internal East group.

Now that we understand how the receptor mesh connects and the details surrounding their deployment, let's learn how to create the inventory to create this deployment.

Inventory creation for the global Automation mesh example

The inventory to create this mesh is fairly straightforward, but there are a lot of parts as it contains seven groups. The peers for each group must be set, as well as the node types. This section will refer to an **INI** inventory example file in this chapter's GitHub repository in the ch14/global_mesh file, though a duplicate of this file in **YAML** format in the redhat_cop.aap_utilities installer is available at ch14/global_mesh_var.yml. They both convey the same information, just in different forms.

The automationcontroller group is important as it is the central hub of this deployment. The group variables set the node type that is set to be controlled as they do not directly interact with the hosts. In addition, the directly connected nodes are peered to this group so that a mesh is formed between these six nodes:

```
[automationcontroller]
control1
control2

[automationcontroller:vars]
peers=instance_group_directconnected
node_type=control
```

The execution_nodes group is all-encompassing. It contains all 12 of the execution and hop nodes. Here, set the node type to execution:

```
[execution_nodes]                    [execution_nodes:vars]
iw_exec1                             node_type=execution
...
```

The directconnected instance group does not get used in practice, but it does contain the four execution nodes from the East and West data centers:

```
[instance_group_directconnected]
iw_exec1
iw_exec2
ie_exec1
ie_exec2
```

Then, two groups have been set: `instance_group_iw_exec` and `instance_group_iw_exec`. Each contains its respective nodes. This is set so they can be assigned jobs:

```
[instance_group_iw_exec]          [instance_group_ie_exec]
iw_exec1                          ie_exec1
iw_exec2                          ie_exec2
```

The next two set the hop group's `dmz_hop` and `use_hop` nodes; `dmz_hop` is peered to the `automation_controller` group directly, and `use_hop` is peered back to `dmz_hop`. Both groups are set to the hop node type:

```
[dmz_hop]                         [use_hop]
dmz1                              chop1
dmz2                              chop2


[dmz_hop:vars]                    [use_hop:vars]
peers=automationcontroller        peers=dmz_hop
node_type=hop                     node_type=hop
```

The final two sets of groups are the `instance_group_uwe_exec` and `instance_group_euc_exec` groups, which are created so that they can be referenced for jobs. Their peer is set to `use_hop`:

```
[instance_group_uwe_exec]
uwe_exec1
uwe_exec2


[instance_group_uwe_exec:vars]
peers=use_hop


[instance_group_euc_exec]
euc_exec1
euc_exec2


[instance_group_euc_exec:vars]
peers=use_hop
```

These groups and variables describe the deployment shown in *Figure 14.4*. This is not going to fit every situation, but pieces of it should be able to be used to fit anyone's deployment that needs to spread execution nodes out to disparate environments.

Summary

This chapter covered how to scale automation using Automation mesh, the details of Automation mesh and how using different types of nodes is important for scaling and growing your automation capabilities.

The next chapter will cover the Automation services catalog, which allows you to abstract jobs, forms, and approvals to a business user-friendly interface.

15

Using Automation Services Catalog

Automation Services Catalog (ASC) is a frontend for users to the Automation controller. It is designed for business users that may not realize that they are using Ansible Automation. It allows users to order jobs and fill in surveys for job templates and workflows through a friendly user interface. This chapter will go into the details of what exactly ASC is, how to configure the various parts of it, how to create orders and products for users to consume, and how to use approval workflows.

In this chapter, we're going to cover the following main topics:

- What is Automation services catalog?
- Configuring ASC
- Approvals inside ASC

Technical requirements

An ASC server is required to follow along with this chapter. Details for creation can be found in *Chapter 2, Installing Ansible Automation Platform.*

What is Automation services catalog?

ASC is a way for business users to have a friendly user interface to order job templates and workflows. This was originally a **Software as a Service (SaaS)** hosted alongside Automation hub on `https://console.redhat.com`. However, due to popular demand, it was moved to be an on-premises solution for use. The upstream product where development work is done is called Pinakes (`https://github.com/ansible/pinakes`), named after the Greek term for library catalogs. For information on how to build an upstream development deployment, refer to the installation guide at the aforementioned Pinakes link.

ASC is made up of the following sections in the GUI:

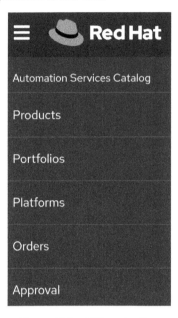

Figure 15.1 – ACS navigation

The navigation bar is the primary way of navigating ASC. The details of each section are as follows:

- **Platforms**: This is the Automation controller that the catalog connects to. At the time of writing, it is limited to connecting to a single Automation controller. It provides information about templates and inventories, as well as a summary of the last sync.

- **Portfolios**: This is a group of products. Each portfolio can be assigned a different approval process. In addition, products within a portfolio have unique attributes such as names, survey customizations, and approvals.

- **Products**: These are job templates and workflows from the Automation controller that can contain survey customizations and an approval process. These can be copied to other portfolios as well.

- **Orders**: This is a list of products ordered similar to jobs in the Automation controller, along with their details.

- **Approval**: Approval processes that portfolios and products can adopt are set here. In addition, pending approvals appear here, waiting for actions to be taken.

These different areas interact with each other. Each of these works in tandem to create orders for users and to have those orders approved to create jobs in the Automation controller. In addition, those jobs can have surveys with additional options and validations done. The following diagram shows how the different parts interact:

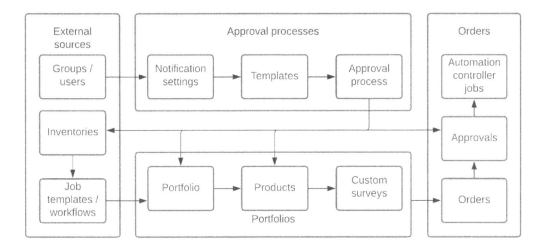

Figure 15.2 – ACS interactions

There are a lot of moving pieces in ASC, so it is good to review the workflow of interactions. Products and inventories originate with the Automation controller, while groups originate with the **Red Hat Single Sign-On** (**RH-SSO**) server. The approval process sets up who can be assigned to approval, and how and when they are notified. The portfolios contain products that can have custom attributes applied, and can then be shared with selected groups. This is all configured by an administrator.

After the configuration is done, the users of the groups can log in, order products, approve orders, and get results, without having to touch the Automation controller. Each of these interactions will be explored in this chapter. The next section will look at how to configure the ASC for use.

Configuring the ASC

This section will go into how you can configure the ASC. This includes SSO configuration and configuring items on the ASC itself. The platform connection is added to the ASC upon installation and does not need to be configured. However, before being able to access the ASC, accounts need to be created. The next section will cover granting groups and users access.

> **Important Note**
>
> The ASC is a new product, so there have not been any collections, roles, or modules built around configuring it. This chapter will go through the GUI and go through every part of the ASC. However, it is recommended to check out `https://github.com/redhat-cop/` `services_catalog_configuration` to see whether a collection has been created to assist in configuration. Such a collection should work similarly to those referenced in other chapters.

Configuring the SSO for access to the ASC

ASC relies solely on a **Single Sign-On (SSO)** server for authentication. All groups and users are pulled from there. Configuring SSO authentication and adding permissions to groups was discussed in *Chapter 4, Configuring Settings and Authentication*; instructions on connecting an identity provider, users, and groups can be found there. Now, let's look at what roles can be assigned to use ASC.

Overview of ASC RBAC

The ASC server has client roles for **Role-Based Access Control (RBAC)** under the name **automation-catalog**, similar to **automation-hub**, which can be assigned in the SSO server to users or groups that have the following permissions:

- `approval-admin`:

  - Create approval groups and assign users to them

  - Have control over the approval processes

  - Monitor, approve, and deny order requests

- `approval-approver`:

  - Monitor, approve, and deny order requests

- `catalog-admin`:

  - Create, edit, and delete order processes

  - Edit portfolios, products, and orders

- `catalog-user`:

  - Create an order request

  - Monitor existing requests

At least one account or group should be set to have all four of these permissions so that they can be the full administrator of ASC. These RBAC roles should be assigned to groups on the SSO server as needed.

If no identity provider is created in SSO, it is recommended to create accounts for use on the SSO server. In the installer, the `sso_automationcatalog_create_user_group` option defaults to `true`. If this option was not turned off, then these sample accounts are created for use. The installer creates three users and groups by default that have access to ASC. The groups, default roles, and users are as follows:

- Group: `Information Technology - Sample`:

  - Roles: `catalog-admin` and `approval-admin`

- User: `fred.sample`
- Group: `Marketing - Sample`:
  - Roles: `catalog-user`
  - User: `barney.sample`
- Group: `Finance - Sample`:
  - Roles: `approval-approver`
  - User: `wilma.sample`

The password for each of these accounts is the user's name – for example, the password for `fred.sample` is `fred`. These accounts and RBACs allow access to ASC to be controlled. In the next section, you will learn how to create approval settings.

> **Important Note**
>
> ASC is a new product being actively developed. To stay ahead of the cycle, several features that we'll discuss will be based on the development version. These will be denoted with **Development setting** and may change by the time they make it to release.

Configuring approval settings

Approval settings can be added to other pieces of ASC. Each part of the approval setting is built on the previous one; they are layered like an onion. They consist of notification settings, templates, and approval processes. These will be covered in order. Each of these is a tab in the **Approval** section of ASC.

Configuring notification settings

Currently a **Development setting**, this allows you to add a way to notify users when events happen. At the time of writing, the only method available for notifications is **Email**, though multiple email settings can be created. To create a new notification, navigate to **Notification Settings**, click **Create**, and fill in the following fields:

- **Name**: The name of the notification item.
- **Notification type**: The type of notification; at the time of writing, this is limited to email.
- **Host**: Hostname of the email server.
- **Port**: Port to use for the email server.
- **Username**: Username to authenticate to the email server.
- **Password**: Password to authenticate to the email server.

- **Security**: SSL/TLS – that is, the security method to use.

- **SSL Key**: The key to use. This can be left blank.

- **SSL Certificate**: The certificate to use. This can be left blank.

- **From**: Email to set the from message to.

Click **Submit** to save this notification. Different types of notifications should be added in the future. These notifications can then be used by templates, which we will cover in the next section.

Configuring templates

Templates are used in the approval process to determine which notification to use. The built-in approval template sends no notifications. To create a template, navigate to **Templates**, click **Create**, and fill in the following fields:

- **Title**: The name of the template.

- **Description**: The description of the template.

- **Process method**: Sends a notification to approvers when a new approval is pending.

- **Signal method**: This sends an acknowledgment to the user when the approval process is finished.

Click **Submit** to save this template. Once a template has been created, it can be used in an approval process, as explained in the next section.

Configuring approval processes

Approval processes are assigned to orders and inventories. This sets what approvals are required for the action to move forward. It contains the following fields:

- **Name**: The name of the process.

- **Description**: The description of the process.

- **Template**: The template to use in the process.

- **Add Groups**: Groups that will approve any request. If multiple groups are added, they will all need to approve the request.

Click **Submit** to save this process. These approval processes are now available for use. Now, let's learn how to configure portfolios.

Configuring portfolios

Portfolios are unique versions of products. To create a portfolio, navigate to **Portfolios**, click **Create**, give it a name and description, and click **Create**. Once created, click on the newly created portfolio, click **Add**, and select the platform (**Automation Controller**) from the drop-down list. From here, click the checkboxes of the workflows and job templates you want as products in that portfolio and click **Add**:

Add products: Execution jobs

Automation Controller   ▼   Filter products   Q   Cancel   Add

1 - 5 of 5   ▾

● **Red Hat** Ansible Automation Platform ☑	● **Red Hat** Ansible Automation Platform ☐	● **Red Hat** Ansible Automation Platform ☐
Simple workflow schema	Simple workflow schema2	test-template-1
a basic workflow	a basic workflow	created by Ansible Playbook

Figure 15.3 – Adding products to a portfolio

Anything that's been added to a portfolio will now show up on the **Products** page of the navigation bar. Products not present in portfolios do not appear. Either on the portfolio page itself or on the page that shows portfolios, an icon with three dots allows you to edit the portfolio:

Order up   1   ⋮

Last updated 2 minut   Share

by   Copy
Jobs ready to orde
                            Set approval

                            Edit

                            Delete

Figure 15.4 – Editing a portfolio

Let's look at these actions in more detail:

- **Share**: This allows you to invite groups to use the portfolio. These groups can be granted permission to either order/edit or order/view products inside portfolios.

- **Copy**: This makes a duplicate portfolio with **Copy of** added to the name that includes any modifications made to the products inside the portfolio.

- **Set approval**: This gives you the option to set one or many of the approval processes for the portfolio.

- **Edit**: Here, you can edit the name and description of the portfolio.

- **Delete**: This option removes the portfolio.

With portfolios of products created, the next step is to edit the products within them.

Configuring products inside portfolios

The **Product** navigation bar page shows all products that the current user can order. If they share the same name, they will show up the same. It is possible to edit the name and description of a product so that it's unique or to see which portfolio it belongs to. You can do this by examining the mouseover **URL** and looking for `portfolio-item?portfolio=3`. You can see which portfolio is which using the same mouseover technique on the portfolio page.

However, for editing a portfolio, it is easier to navigate to the portfolio page and edit the products from within there. It is recommended to add the portfolio name as either a prefix to the product name or in the description so that users know which one they are ordering from. It is also possible to entirely change the product name so that it's different from the workflow or job template. I am a fan of naming the job templates and workflows with descriptive names already, so I use the description field to differentiate them.

On the portfolio page, if you click the name of a product, it will take you to a page that has the options for **Order** and the three dots icon for making changes. These options are as follows:

- **Order**: Order a product. If there is a survey attached, it will prompt the user to fill it out. If there is an approval process attached, it will start that process. Ordering will be covered in more detail in the next section.

- **Edit**: Change the name or description of the order.

- **Copy**: Make a copy of the order to any portfolio you can edit. If the copy goes to the same portfolio, **Copy of** will be added to the name.

- **Set approval**: Here, you can set one or many of the approval processes for the product. This is set at the product level, regardless of anything set at the portfolio level.

- **Edit survey**: This allows you to make changes and add validation to the survey if one is enabled on the Automation controller.

Survey edits stay with the product itself, not the underlying template/workflow it's connected to. Now, let's learn how to customize surveys.

Survey changes and validation

Survey customization is one of the benefits of ASC. It consists of setting properties and validations. The screen for this looks as follows:

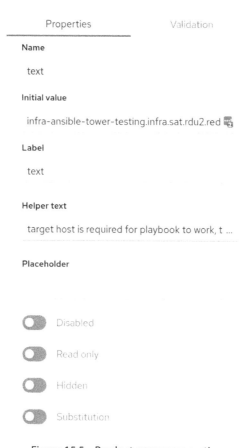

Figure 15.5 – Product survey properties

Each survey entry has the following effect:

- **Initial value**: Change the default value of the field.
- **Label**: Change the question's wording.
- **Helper text**: Change the description.
- **Placeholder**: This creates a grayed-out value when the user has not entered a value. It defaults to the initial value and is used to suggest a value for the user to use.

For each of these values, check the documentation as their features are pending. At the time of writing, this is what they do:

- **Disabled**: Prevents the user from making changes. This should change in the future.
- **Read Only**: Prevents the user from making changes.
- **Hidden**: The user cannot see the question.
- **Substitution**: Enables the disabled option.

These can be set for each survey entry. In addition to these properties, validation can be performed, as shown in the following screenshot:

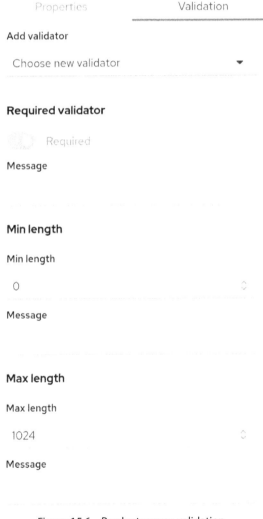

Figure 15.6 – Product survey validation

Not all the validators have been shown in the preceding screenshot; more can be added for validation. Each of these has an option such as length or pattern, and a message to send to the user when it fails validation. **Min length**, **Max length**, **Exact length**, **Min items**, **Min number value**, and **Max number value** are the self-explanatory validations that can be made.

The following validators can also be used:

- **Pattern**: Specifies a regular expression that defines a pattern the survey entry needs to follow
- **URL**: Validates a variety of URL options, including the protocol, host, port, IPv4, IPv6, and other URL validation options

These validators can also be configured to send messages to users when the validation fails:

Figure 15.7 – Validation failure

These tools, coupled with the GUI and approvals, allow for some nuance and checks before orders are sent to the Automation controller. Now, let's learn how to use approvals in ASC.

Approvals inside ASC

Approval processes can be applied to multiple places. In the GUI, the following areas inside ASC allow for approvals to be set:

- **Use of an inventory**: **Platform**, **Platform name**, **Inventories**, **Inventory name**, **Set approval**, **To set approval for an Automation controller inventory**
- **All orders in a portfolio**: **Portfolio**, **Portfolio name**, **Set approval**
- **Use of a product**: **Products**, **Product name**, **Set approval**

An approval process can be set at each of these levels. If a group is designated in multiple approval processes, the group will only be prompted for approval once. These are processed as individual requests but can be seen on the **Approval | All requests** page:

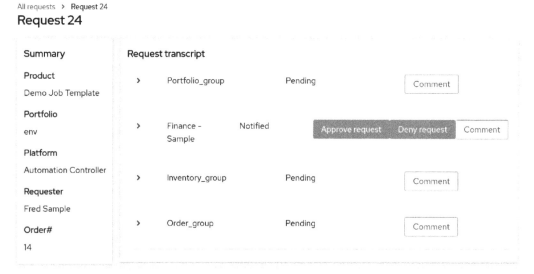

Figure 15.8 – Approval requests

For illustration purposes, I've created groups and processes that match the area that requires approval. Each group will need to approve this order before it is sent to the Automation controller. Approvals do not need to be set at every level; it can make sense in some cases, but it is good that there are options to set them.

Summary

This chapter covered ASC, including how to create approvals and the notifications for the approval processes, how to edit and validate surveys for users to use in orders, and how to create and customize portfolios and products.

For some users and use cases, this will be a better interface to create jobs from job templates than the Automation controller. For other, more advanced users, it may not be. ASC may not be the best tool in every use case, but it does provide a large number of knobs and tools that you can tweak, in addition to those in the Automation controller, and it is a welcome addition to Ansible Automation Platform.

Afterword

You should be well on your way to better utilizing Ansible Automation Platform, and utilizing **Configuration as Code** (**CaC**). Thank you for making it to the end of this book – I hope you enjoyed reading it as much as I enjoyed writing it. If you have found any issues, corrections, suggestions, or have any other feedback, feel free to open an issue in this book's repository! I will strive to be present there and to maintain and update the code provided so that it continues to be useful, working, and accurate.

Index

`Packt.com`

Subscribe to our online digital library for full access to over 7,000 books and videos, as well as industry leading tools to help you plan your personal development and advance your career. For more information, please visit our website.

Why subscribe?

- Spend less time learning and more time coding with practical eBooks and Videos from over 4,000 industry professionals

- Improve your learning with Skill Plans built especially for you

- Get a free eBook or video every month

- Fully searchable for easy access to vital information

- Copy and paste, print, and bookmark content

Did you know that Packt offers eBook versions of every book published, with PDF and ePub files available? You can upgrade to the eBook version at `packt.com` and as a print book customer, you are entitled to a discount on the eBook copy. Get in touch with us at `customercare@packtpub.com` for more details.

At `www.packt.com`, you can also read a collection of free technical articles, sign up for a range of free newsletters, and receive exclusive discounts and offers on Packt books and eBooks.

Other Books You May Enjoy

If you enjoyed this book, you may be interested in these other books by Packt:

Practical Ansible 2

Daniel Oh, James Freeman, Fabio Alessandro Locati

ISBN: 9781789807462

- Become familiar with the fundamentals of the Ansible framework
- Set up role-based variables and dependencies
- Avoid common mistakes and pitfalls when writing automation code in Ansible
- Extend Ansible by developing your own modules and plugins
- Contribute to the Ansible project by submitting your own code
- Follow best practices for working with cloud environment inventories
- Troubleshoot issues triggered during Ansible playbook runs

Mastering Ansible - Fourth Edition

James Freeman, Jesse Keating

ISBN: 9781801818780

- Gain an in-depth understanding of how Ansible works under the hood
- Get to grips with Ansible collections and how they are changing and shaping the future of Ansible
- Fully automate the Ansible playbook executions with encrypted data
- Use blocks to construct failure recovery or cleanup
- Explore the playbook debugger and Ansible console
- Troubleshoot unexpected behavior effectively
- Work with cloud infrastructure providers and container systems

Packt is searching for authors like you

If you're interested in becoming an author for Packt, please visit `authors.packtpub.com` and apply today. We have worked with thousands of developers and tech professionals, just like you, to help them share their insight with the global tech community. You can make a general application, apply for a specific hot topic that we are recruiting an author for, or submit your own idea.

Share your thoughts

Now you've finished *Demystifying Ansible Automation Platform*, we'd love to hear your thoughts! Scan the QR code below to go straight to the Amazon review page for this book and share your feedback or leave a review on the site that you purchased it from.

https://packt.link/r/1803244887

Your review is important to us and the tech community and will help us make sure we're delivering excellent quality content.

www.ingramcontent.com/pod-product-compliance
Lightning Source LLC
Chambersburg PA
CBHW062107050326
40690CB00016B/3239